Sustainable Civil Infrastructures

About this Series

Sustainable Infrastructure impacts our well-being and day-to-day lives. The infrastructures we are building today will shape our lives tomorrow. The complex and diverse nature of the impacts due to weather extremes on transportation and civil infrastructures can be seen in our roadways, bridges, and buildings. Extreme summer temperatures, droughts, flash floods, and rising numbers of freeze-thaw cycles pose challenges for civil infrastructure and can endanger public safety. We constantly hear how civil infrastructures need constant attention, preservation, and upgrading. Such improvements and developments would obviously benefit from our desired book series that provide sustainable engineering materials and designs. The economic impact is huge and much research has been conducted worldwide. The future holds many opportunities, not only for researchers in a given country, but also for the worldwide field engineers who apply and implement these technologies. We believe that no approach can succeed if it does not unite the efforts of various engineering disciplines from all over the world under one umbrella to offer a beacon of modern solutions to the global infrastructure. Experts from the various engineering disciplines around the globe will participate in this series, including: Geotechnical, Geological, Geoscience, Petroleum, Structural, Transportation, Bridge, Infrastructure, Energy, Architectural, Chemical and Materials, and other related Engineering disciplines.

More information about this series at http://www.springer.com/series/15140

Zahid Hossain · Jiupeng Zhang
Can Chen
Editors

Solving Pavement and Construction Materials Problems with Innovative and Cutting-edge Technologies

Proceedings of the 5th GeoChina International Conference 2018 – Civil Infrastructures Confronting Severe Weathers and Climate Changes: From Failure to Sustainability, held on July 23 to 25, 2018 in HangZhou, China

Springer

Editors
Zahid Hossain
Department of Civil Engineering
Arkansas State University
Jonesboro, AR, USA

Can Chen
ChemCo Systems
Redwood City, CA, USA

Jiupeng Zhang
School of Highway
Chang'an University
Xi'an, China

ISSN 2366-3405 ISSN 2366-3413 (electronic)
Sustainable Civil Infrastructures
ISBN 978-3-319-95791-3 ISBN 978-3-319-95792-0 (eBook)
https://doi.org/10.1007/978-3-319-95792-0

Library of Congress Control Number: 2018948637

This Springer imprint is published by the registered company Springer Nature Switzerland AG
The registered company address is: Gewerbestrasse 11, 6330 Cham, Switzerland

Contents

Introduction

Extreme weather events and climate conditions pose challenges for civil infrastructures, and they are believed to cause premature distresses, endanger public safety, and cost a huge amount of taxpayers' money. Selection and design proper are considered prerequisites for durable and longer lasting infrastructures construction materials and can mitigate the aforementioned infrastructure problems. In recent years, construction engineering professionals around the globe have strived to solve critical infrastructure problems. This volume of the conference proceedings presents findings of research involving topics such as soft subgrade soils, poor quality aggregates, deformation of asphalt pavements, thermal and fatigue cracking of concrete pavements, the stability of slopes, surface irregularity of airport pavements, and challenges with freight movements. Research findings presented in the proceedings suggest that each infrastructure problem is unique in the contexts of its location, purpose, and service conditions. However, a viable solution to the problem requires combined efforts of design engineers and practitioners around the world as the future depends on the worldwide professionals who invent, apply, and implement these technologies. Results from innovative and cutting-edge experimental works and data from in-service conditions of these infrastructure problems presented here are expected to aid world professionals to learn from these examples and solve construction problems of the twenty-first century.

Finding an Optimal Bitumen and Natural Sand Balance for Hot Mix Asphalt Concrete in Hot and Arid Regions

Fathi S. Almadwi(✉) and Gabriel J. Assaf

Department of Civil Engineering and Construction, École de Technologie Supérieure (ÉTS), University of Québec, 1100 Rue Notre-Dame Ouest, Montréal, QC H3C, Canada
fathi.madwi.1@ens.etsmtl.ca, Gabriel.Assaf@etsmtl.ca

Abstract. Designing a Hot Mix Asphalt Concrete (HMAC) for Libya's southern desert regions has two serious challenges. Due to the extremely hot and arid conditions, where road surfaces can be reaching 70 °C, and with air humidity well below 50%, a HMAC in the southern desert often suffers great deformations in the form of rutting. This is particularly a problem in lower-volume roads because the remoteness and the low traffic of these roads makes an investment in their construction and repair not cost effective to use higher-grade materials. This paper shows comparative data for paving mix performance of two bitumen grades available in Libya, both under simulated climatic conditions. These are used to select the best mix of sand and bitumen. The research first determines the best mix of natural desert sand with manufactured aggregate. Then it uses this sand mix as a control to evaluate two kinds of bitumen, Bitumen 60/70 (B60/70) and PG70-10. With a fine aggregate mix of 33% natural desert sand (from the region) and 63% manufactured crush fine aggregate, the performance of PG70-10 was found to be superior compared with B60/70. These are assessed with the Marshall test, the Super Gyratory Compactor test, and the wheel track test in the ETS Laboratories (École de technologie supérieure) Faculty of Engineering of the University of Québec, Canada. This finding is important for designers working in arid conditions as a way of sizable cost savings by substituting available materials, i.e. the rounded sands of desert regions for more expensive sands imported from other regions.

Keywords: Natural sand · Hot mix asphalt · PG70-10 · B60/70
Fine aggregate · Libya

1 Introduction

The roads in the desert areas of southern Libya are constrained by two factors. They must be built on a strict budget because of their low-volume use but they suffer from extreme temperatures in the arid conditions and under the hot desert sun. For this reason, the roads are typified by excessive deformation (mostly rutting and some cracking). These roads are made with Bitumen 60/70 (B60/70) because the bitumen is an established formulation and widely available from local refineries. Due to the

Z. Hossain et al. (eds.), *Solving Pavement and Construction Materials Problems with Innovative and Cutting-edge Technologies*, Sustainable Civil Infrastructures,
https://doi.org/10.1007/978-3-319-95792-0_1

extreme conditions, it performs poorly in extreme heat and the result is a large degree of rutting in the Hot Mix Asphalt Concrete (HMAC). At their most extreme, the deserts in the south of the country attain daytime air maximum temperatures of 55 °C, which causes road surface temperatures to go over 70 °C (Almadwi and Assaf 2017).

2 Scope and Objective of Study

In areas with arid sand deserts, like southern Libya, there are practical reasons to use desert sand in asphalt mix but too much natural sand in HAMC results in an HMAC more likely to show stresses leading to permanent deformation in use. This paper aims to demonstrate the effects of natural sand on the in-service use of asphalt mixture. Here two objectives are sought. The first is to find the optimal balance of natural desert sand with manufactured fine aggregate; the second is to select the better bitumen formulation for the conditions. Once the mix of sand is determined, the two bitumen formulations are mixed, and the different mixes undergo various tests including the Marshall, Gyration, and Wheel Track tests. Results from the tests are used to evaluate the different mixes with the same equipment in order to predict the behaviour of the asphalt mixtures with the materials. Results are analyzed to determine the effects of the different factors in the behaviour of the asphalt.

3 Background

3.1 Material Choice and How It Affects HMAC Properties

Every component in the HMAC has an effect on its final characteristics and how these respond to moisture, fatigue leading to cracking and permanent deformation. Other factors, that need to be present in testing in order to properly simulate actual conditions are loading distribution and climate (Epifanio and Gan 2009).

3.1.1 Aggregate

The quality and consistency of aggregate is an important factor in pavement performance. More specifically, in order to resist permanent pavement deformations, the durability, angularity, and chemical composition of an aggregate must be considered. In certain regions with very high temperatures, concrete mixes can incorporate partially crushed aggregate or aggregates containing natural desert sand. 94–95% of a HMAC is made up, by weight, of aggregates and these aggregates create an effective structural matrix that carries most of the weight of the traffic load. Therefore, aggregate selection is a key stage in designing a quality pavement that will be durable over time (Asphalt Institute 1998). That is to say, that, when planning a pavement for a long service life (i.e. a pavement that is not prone to permanent deformation) it is the matrix structure of the aggregate that must be kept in mind. The structural matrix that the aggregate creates is designed to receive and dissipate the energy from years of axle loads without itself changing its overall structure. In creating this structural matrix, individual pieces of fine aggregate must be considered for their coefficient of friction (i.e. texture), their form, as well as the gradation of the whole. When these pieces of aggregate have greater coefficients of friction between pieces and have greater angularly in form, the resulting structural matrix will have greater sheer strength than a HMAC made with smooth, and rounded aggregate. Table 1 shown some useful numbers of coefficient of friction on the engineering properties of material. A structural matrix made with aggregate that has a high coefficient of friction and that has more angular, more cubical pieces will be more stable because the aggregate will interlock and work together as an elastic whole; in contrast, a HMAC made from smooth, rounded aggregate will not interlock as much and pieces of aggregate will start to migrate in relation to each other and the matrix will be less stable in the long-term (Button 1991; Chowdhury et al. 2001; Park and Lee 2002).

Table 1. Coefficient of angle friction (ϕ) on the engineering properties of material

Angle of internal friction (ϕ)	
Rock	30°
Sand	30–40°
Gravel	35°
Silt	34°
Clay	20°
Gravel with some sand	34–48°
Silt	26–35°

3.1.2 Asphalt Binder

The quality of the bitumen, with regards to the aggregate type, will affect the flexibility, stability, fatigue resistance, and durability of the asphalt mix, as well as how it resists moisture damage (Hay and Kopack 1986). Too much bitumen leads to bleeding under traffic loads and poor pavement performance. The properties of the hot asphalt mix are affected by the physical properties of the bitumen (Roberts et al. 1991). The conventional B60/70 penetration grade bitumen, commonly used in road projects in hot areas. For reference, the hardness of the bitumen is evaluated based on how far a specified needle will vertically penetrate the sample within five seconds at 25 °C. Therefore, the softer bitumen is the higher penetration number. The number is shown as a range to the tenth of a millimeter; therefore B60/70 would indicate that the needle penetrated to an average range of between 0.6 and 0.7 mm. The softening point of B60/70 is 49–54 °C which is below daytime surface temperatures of pavement in Libya, where they often reach 70 °C. Options to solve this problem include either a polymer-modified asphalt or a higher, harder grade of bitumen. This project investigates a higher quality of bitumen (Almadwi and Assaf 2017).

4 Methodology

4.1 Materials and Testing Procedures

4.1.1 Aggregate

This study was conducted using dry samples obtained from the sand dune area in the southwestern desert of Libya, around the city of Sabha. Sand from the region was collected and subjected to identification to determine its engineering properties, including sieve analysis, specific gravity and absorption. The size of the manufactured sand in the aggregate mixture was 0–5 mm. The volumetric properties for the natural Libyan sand and the manufactured sand is shown in Table 2. Figure 1 shows the grain size distribution for both. In this research, a mineral filler powder from a Canadian manufacturer was used in small quantities (4%) and combined with asphalt sand mixes; the resulting properties are shown in Table 2.

Table 2. Aggregate properties

Aggregate	Apparent specific gravity	Bulk specific gravity	(%) Absorption
Natural sand 0–5 mm	2.63	2.42	0.33
Manufactured sand 0–5 mm	2.639	2.44	0.58
Mineral filler	2.70	–	1

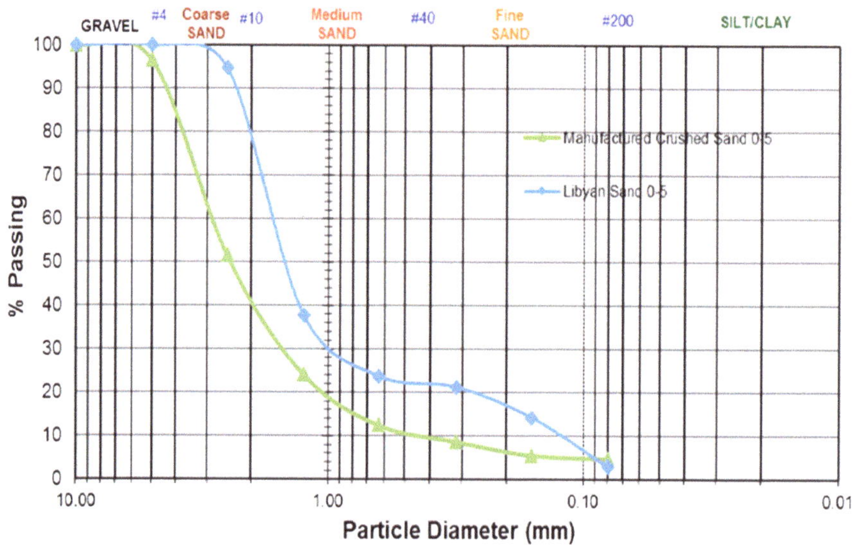

Fig. 1. Grain size distribution of fine aggregates

4.1.2 Asphalt Binder

Due to the conditions discussed in the background, the solution chosen here is to use bitumen PG70-10 and B60/70. Their recommended temperatures are as follows: PG70-10 is recommended to be mixed at 162 °C and compacted at 154 °C. B60/70 is recommended to be mixed at 156 °C and compacted at 143 °C. The recommended temperatures were followed without exception, as illustrated in Table 3. The dynamic shear rheometer (DSR) is used to characterize the viscous and elastic behavior of asphalt binders at medium to high temperatures. This characterization is used in the Superpave PG asphalt binder specification. As with other Superpave binder tests, the actual temperatures anticipated in the area where the asphalt binder will be placed determine the test temperatures used. Figure 2 shows the Brookfield viscosity as compared to temperature equivalents and how they relate to the binders used here.

Table 3. Properties of asphalt binders and temperatures for mixing and compacting

Asphalt binders	Bulk specific gravity	Mixing temperature (°C)	Compacting temperature (°C)
PG70-10	1.29	162	154
B60/70	1.02	156	143

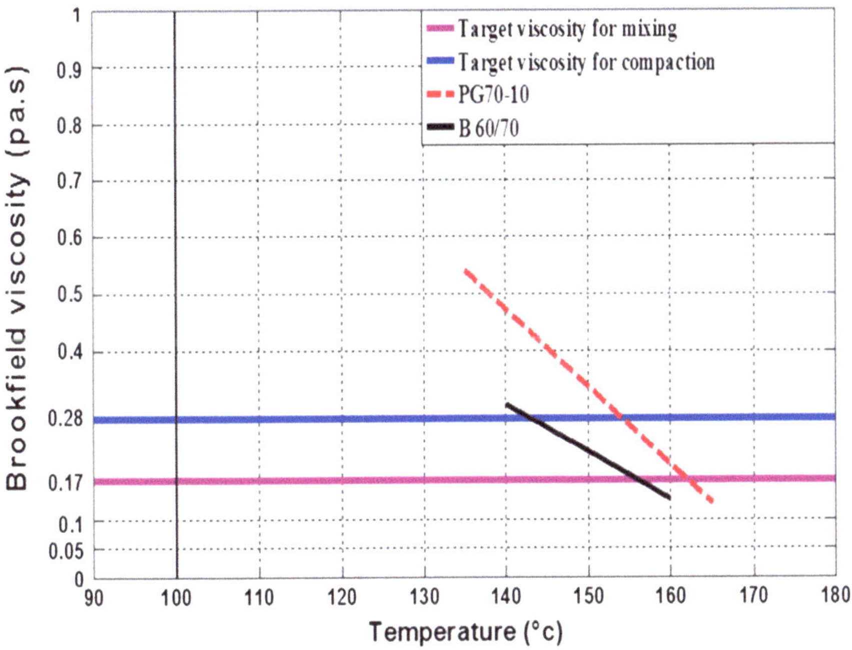

Fig. 2. Brookfield temperature versus viscosity curve for asphalt (Almadwi and Assaf 2017)

4.2 Determining the Optimal Natural Sand Mix

Several types of testing equipment and test procedures were used to determine the amount of natural sand required. For all of these tests, both PG70-10 and B60/70 were used. The tests were carried out with 50% natural sand content, but this mix was found to be insufficient in terms of strength and stability. Following tests were carried out with 25% natural sand content and these showed that the amount of natural sand was not a problem in terms of the final mix quality. Nonetheless, the objectives of the project are to find a workable maximum amount of natural sand; therefore, more tests were done. Following tests used 30 and 35%. At 30%, the resulting pavement showed good characteristics but at 35% the tests showed too much deformation. As a result, a natural sand content of 33% by weight was selected for the samples in this study with regard to the total mass of the aggregate mixture.

4.3 Determining the Optimal Bitumen Content

Optimal bitumen content (OBC) was determined according to the Marshall method. To determine the OBC, the tests were carried out with 4.5% bitumen content, but these were found to be insufficient to bind the mix. Following tests were carried out with 5% bitumen content and these showed that the amount of bitumen was not sufficient. Following tests used 5.5 and 6%. The test with 6% bitumen resulted in too much bitumen bleeding. As a result, bitumen content of 5.5% by weight was selected which corresponded to 4% air voids that were fulfilled the requirements of Marshall method for the samples in the following tests.

4.4 Test Procedure

In Libya, hot asphalt mixtures have been prepared according to the Marshall method, which does not take into consideration the shear strength during the mix design phase. For this reason, it does not select out the mixes that perform poorly with regards to rutting. Present mix design methods, according to Mahboub (1990), have gone beyond the scope of what they were meant to do because road use has changed since they were first designed. Examples of differences are that tire pressure, axle-load and environmental conditions are different from what they were even a decade ago. The approach for making the aggregate mix consisted of adding a percentage of natural sand to manufactured sand, 33 and 63%, respectively, by weight of the mix. About 4% of mineral filler was also added.

4.5 Laboratory Tests for HMAC Design

Manufactured aggregate, natural sand, and filler, all between 0 and 5 mm, were used in the laboratory study. The specification of the aggregate gradation for hot mix design is illustrated in Table 4. The natural sand was imported from the desert regions of southern Libya; the bitumen for the tests were PG70-10 and B60/70 grades. Figure 3 shows the specification of aggregate gradations for hot mix design.

Table 4. Fine aggregate grain size distribution for hot mix design

Fine aggregate grain size distribution for hot mix design					
Agg size	L. limit	U. limit	% Retained	% Passing	% Cumulative retained
14	100	100	0.00	100	0
10	95	100	2.54	97.5	2.54
5	85	100	4.68	95.6	7.22
2.5	50	70	32.25	65.4	39.47
1.25	24	34	39.99	29.6	79.46
0.63	16	24	13.49	17.4	92.95
0.315	14	22	3.96	14.1	96.91
0.16	8	12	1.94	9.8	98.85
0.08	4	8	0.83	5.4	99.68
Pan			0.32		100.00

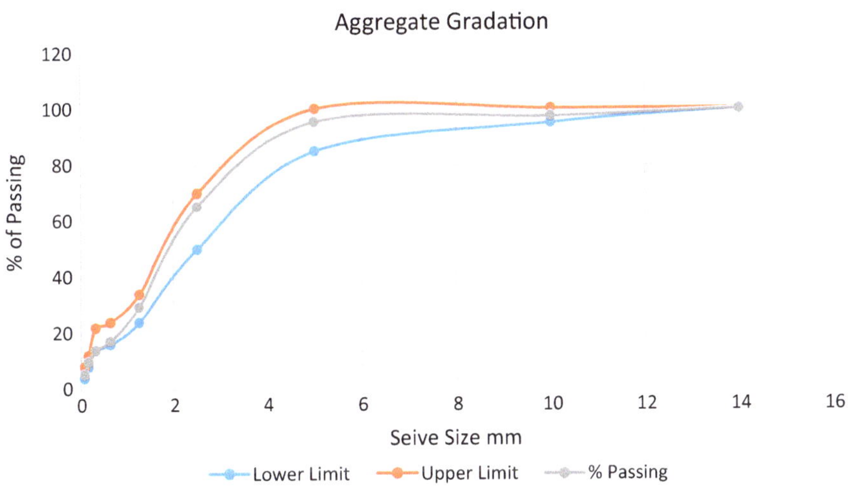

Fig. 3. Specification of aggregate gradation for hot mix design

4.5.1 Marshall Stability Test

Test specimens were prepared by applying 50 blows to the top and to the bottom of the samples with a Marshall hammer. Of the samples, half-used PG70-10 binder and the other half-used B60/70 binder. The tests determined the following volumetric properties of the samples: the percentage of voids filled with asphalt (VFA); the stability and plastic flow; the percentage of voids in mineral aggregate (VMA); the percentage of air voids (Va); the Optimum Bitumen Content (OBC) of the mix.

4.5.2 SUPERPAVE Gyratory Compaction Test

The compacted specimens produced by these tests provided data on HMAC at each gyration of the compaction procedure. Samples were measured at the following intervals: 10, 80, 100 and 200 gyrations. PG70-10 binder samples were compacted at 154 °C; the B60/70 binder samples were compacted at 143 °C.

4.5.3 Wheel Track Test

The test samples were subject to a loaded wheel test to simulate the passage of traffic over time. Binders with different softening points made it necessary to adjust the machine temperature; likewise, tire pressure and the weight of the wheel were adjusted to simulate different axle loads. The first test specified 1000 cycles at 600 kPa tire pressure (at standard room temperatures of ~20°C); tests were then carried out at 1000, 3000, 10,000 and 30,000 cycles (cumulatively) at 65 °C. After each stage, the rutting depth was measured. Following the standards set by the Ministère des Transports de Quebec (2016), the limit for how much the sample can lose from its original height (in mm), following a certain number of heated cycles (cumulative) at 65 °C for the samples is 10% at 1000 heated cycles and 15% at 3000 heated cycles.

5 Results

5.1 Results of the Marshall

The results of the Marshall tests are seen in Fig. 4. For the Volume of Mineral Aggregate (VMA), the results are as follows. VMA with PG70-10 there were two samples: the first sample (S1) resulted in 14.14; the second sample (S2) resulted in 14.26. VMA with Bitumen 60/70 there were also two samples: the first sample (S3) resulted in 16.52; the second sample (S4) resulted in 16.63. For the Void Filled Asphalt (VFA), the results were as follows. VFA with PG70-10 there were two samples: the first sample (S1) resulted in 78.3; the second sample (S2) resulted in 77.5. VFA with Bitumen 60/70 there were also two samples: the first sample (S3) resulted in 63.19; the

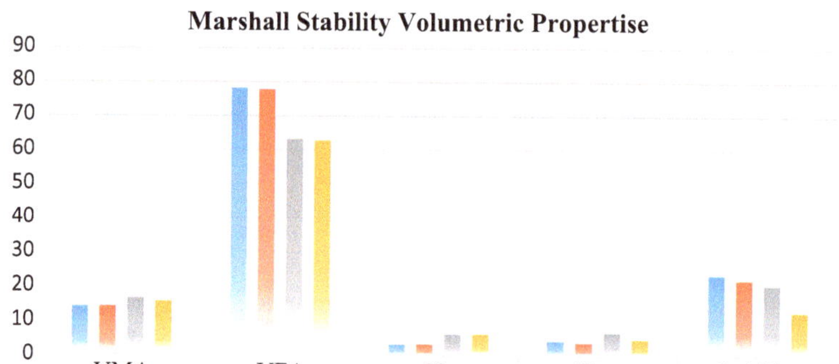

	VMA	VFA	Va	Flow	Stability
PG 70-10 S1	14.14	78.3	3.08	4	23.654
PG 70-10 S2	14.26	77.5	3.21	3.5	21.812
B 60/70 S3	16.63	63.19	6.08	6.4	20.285
B 60/70 S4	15.59	62.69	6.21	4.4	12.82

Fig. 4. Marshall stability results

second sample (S4) resulted in 62.69. For the Air Void percentage (Va%), the results were as follows. Va% with PG70-10 there were two samples: the first sample (S1) resulted in 3.08; the second sample (S2) resulted in 3.21. Va% with Bitumen 60/70 there were also two samples: the first sample (S3) resulted in 6.08; the second sample (S4) resulted in 6.21. For the Marshall Flow tests, the results were as follows. Flow with PG70-10 there were two samples: the first sample (S1) resulted in 4; the second sample (S2) resulted in 3.5. Flow with Bitumen 60/70 there were also two samples: the first sample (S3) resulted in 6.4; the second sample (S4) resulted in 4.4. For the Stability test, with PG70-10 there were two samples: the first sample (S1) resulted in 23.654; the second sample (S2) resulted in 21.812. VMA with Bitumen 60/70 there were also two samples: the first sample (S3) resulted in 20.285; the second sample (S4) resulted in 12.82.

5.2 Results of the SUPERPAVE Gyratory Compactor (SGC)

The SGC tests were done with PG70-10 on two samples, and done on B60/70, also with two samples. See Fig. 5.

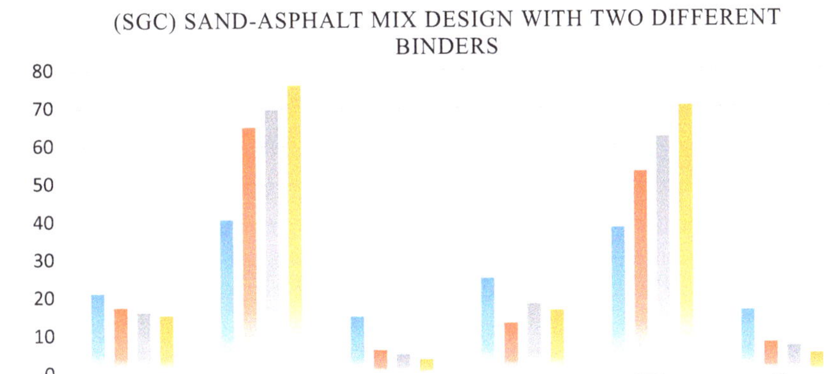

No of Gyration	VMA	VFA (PG 70-10)	Va	VMA	VFA (B 60/70)	Va
10	21.4	40.815	15.03	25.095	38.365	16.393
80	17.61	65.055	6.16	13.06	53.093	7.815
100	16.305	69.645	5.075	18.24	62.28	6.885
200	15.435	76.135	3.665	16.485	70.46	4.88

Fig. 5. Superpave gyratory compactor results

5.3 Results of the Wheel Track

As shown in Fig. 6, samples containing PG70-10 showed less rutting than samples containing B 60/70 at all stages of the rutting test (1000–30,000 cycles cumulative). At 3000 cycles, the PG70-10 samples lost an average of 4.39% of their original height; in contrast, after 3000 cycles, the B60/70 samples lost an average of 8.76% of their original height.

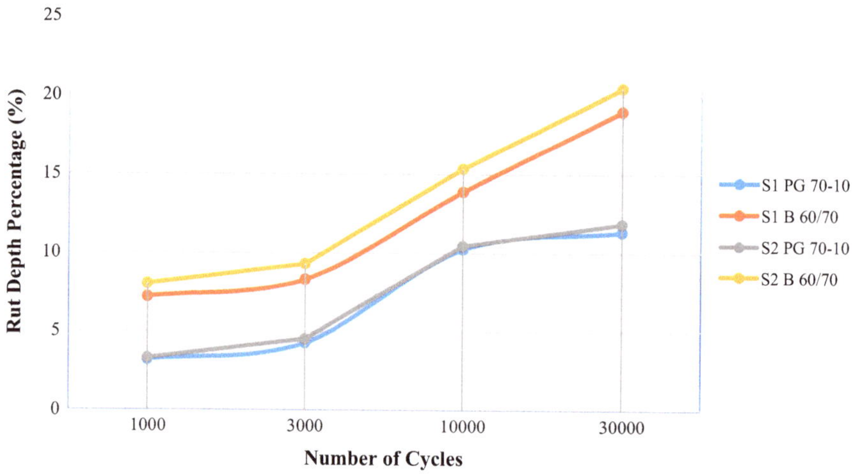

Fig. 6. Wheel track results

6 Discussion

The first step was to determine the amount of natural sand required for all the following tests (with the bitumen). The first tests to determine the natural sand percentage were carried out with 50% natural sand content, but this gave a poor strength and stability. Following tests were carried out with 25% natural sand content and these tests gave a very good result but the project objectives were not to only find a good asphalt pavement but to make use of the available natural sand in the south of Libya, so a higher mix of natural sand was preferred; therefore, more tests were done. Following tests using 30 and 35% were done. The results of the tests with a natural sand aggregate percentage of 30% were good; the same test at 35% natural sand showed too much deformation in the HMAC. Following these tests, a 33% natural sand aggregate mix (by weight) was chosen for both HAMC for this study.

The three tests (Marshall, Gyratory Compactor, Wheel Track) indicate that the mixture of natural sand and PG70-10 performed better than a mixture of natural sand and B60/70 overall. The tests found that the mixture of natural sand and PG70-10 was generally more suitable than the mixture of natural sand and B60/70 for the hot and arid conditions under consideration.

For all the Marshall tests a fine aggregate mix of 33% natural sand and 63% manufactured aggregate was used as a control. As shown in Fig. 6, the Marshall tests samples using PG70-10 all had acceptable results. The tests using B60/70 had acceptable results except for the Marshall Flow test. For this reason, PG70-10 is preferred. Also, the percentage of Air Voids (Va) in the mixture of natural sand and PG70-10 was within the acceptable limit of 4%, but the Va% of natural sand and B60/70 exceeded this recommendation. For both bitumen types, in the VMA, VFA, Va %, and Stability tests; these were all within tolerances. The Marshall Flow of the natural sand and PG70-10 mix was within the target range of 2-4 mm, but the natural sand and B60/70 Marshall Flow mix exceeded this limit according to the MTQ standard (Ministère des Transports de Quebec 2016).

The Gyratory tests were shown in Fig. 5 and illustrated that samples that contained natural sand and PG70-10 had a smaller percentage of air voids than the samples that contained natural sand and B60/70. This was true at all gyration counts. Most importantly, the samples that contained natural sand and PG70-10 had air void content within the 4–7% target range for N-des at 80 and 100 gyrations. The point that both the samples with Bitumen 60/70 exceeded the test criteria was at 80 gyrations. The other test that exceeded the tolerances was the VMA test; with the samples with Bitumen 60/70, the test for VMA exceeded the criteria set for the tests. All other tests were within tolerances (Ministère des Transports de Quebec 2016).

In the Wheel Track test, as displayed in Fig. 6, the slope of natural sand with PG70-10 changed noticeably after 10,000 cycles. From 1000 to 10,000 cycles, the slope of the natural sand with PG70-10 was comparable to the slope of natural sand with B60/70. But from 10,000 to 30,000 the test results from the natural sand with PG70-10 have a much lower slope. This is where the slope of natural sand with PG70-10 contrasts the most with the slope of the natural sand with the B60/70 results. This has important implications for a road surface in actual use. Where the natural sand with B60/70 undergoes considerable ongoing compaction, the natural sand with PG70-10 shows much more general stability over time (Ministère des Transports de Quebec 2016).

7 Conclusions

Natural desert sand is much more worn down and therefore less angular than comparably sized manufactured aggregates; therefore, the functional characteristics of the resulting asphalt mixes are very different. The factor governing resistance to deformation in asphalt mixes are twofold: the capacity to resist shearing between aggregate and bitumen and the percentage of air voids in the bitumen mix. For these reasons, this project determined the highest practical percentage of natural desert sands that could be used in a quality HAMC for low-volume roads in high temperature remote regions of the Libyan desert; this was found to be 33%. The other factor affecting the general resistance to permanent deformation was the choice of the bitumen, necessary to create a bond between the particles. This project determined the better of two bitumen formulations, when this mix of desert sand was necessary. The recommended bitumen is PG70-10.

8 Further Work

Further research plans to investigate the performance result of different additives in the asphalt mixes. Such additives could include polymers or fibers. The information gained by these studies would provide data resources to engineers planning the next generation of road construction in hot and arid countries. The result will be the construction of better-performing roads with lower maintenance costs.

Acknowledgements. The first author wishes to thank the Department of Civil and Construction Engineering at (ETS), for providing the laboratory equipment for this research.

Reference

Almadwi, F.S., Assaf, G.J.: Performance testing of paving mixes for Libya's hot and arid conditions, using marshall stability and SUPERPAVE gyratory compactor methods. In: International Congress and Exhibition Sustainable Civil Infrastructures: Innovative Infrastructure Geotechnology, pp. 313–323. Springer, Cham (2017, July)

Asphalt Institute: Construction of Hot Mix Asphalt Pavement (MS-22), 2 edn. Lexington, Kentucky, USA (1998)

Chowdhury, A., Button, J.W., Grau, J.D.: Effects of Superpave restricted zone on permanent deformation (2001)

Epifanio, E.G., Gan, L. Validation of a Pavement Performance Model for Flexible Pavements Based on (2009)

Hay, R.E., Kopac, P.A.: Relationship of quality control to pavement performance. In: Solutions for Pavement Rehabilitation Problems (1986)

Mahboub, K.: Asphalt concrete creep as related to rutting. J. Mater. Civ. Eng. **2**(3), 147–163 (1990)

Ministère des Transports (Quebec): Normel LC26. VII 4202, 1–17 (2016)

Park, D.W., Lee, H.S.: Test methods for fine aggregate angularity considering resistance of rutting. KSCE J. Civ. Eng. **6**(4), 421–427 (2002)

Perdomo, D., Button, J.W.: Identifying and Correcting Rut-Susceptible Asphalt Mixtures. Final report (No. FHWA/TX-91/1121–2F) (1991)

Roberts, F.L., Kandhal, P.S., Brown, E.R., Lee, D.Y., Kennedy, T.W.: Hot Mix Asphalt Materials, Mixture Design and Construction, 2nd edn. National Center for Asphalt Technology, NAPA Education Foundation, Baltimore, Maryland (1991 & 1996)

Incremental Rutting Prediction with Asphalt Mixture Shear Properties

Josef Zak[1(✉)], Erdem Coleri[2], and John Harvey[3]

[1] Faculty of Civil Engineering, CTU in Prague, Thakurova 7, Prague Czech Republic
josef.zak@fsv.cvut.cz

[2] School of Civil and Construction Engineering, Oregon State University, 101 Kearney Hall, Corvallis, OR 97331, USA

[3] University of California Pavement Research Center, Department of Civil and Environmental Engineering, University of California, Davis, Davis, CA 95616, USA

Abstract. It is a purpose of the article to present the correlation between two devices developed to measure shear properties of asphalt mixtures. The shear properties are measured by well-known Superpave Shear Tester and developed Uniaxial Shear Tester. The material characteristics are determined from two test methodologies. Shear small amplitude oscillation test, known also as a frequency sweep shear test and repeated shear tests are utilized. The properties, characterizing the complex moduli, are determined with the use of sigmoidal function used in Mechanistic Empirical Pavement Design Guide (Hallin in NCHRP 1-37A guide for mechanistic-empirical design. National Cooperative Highway Research Program, 2004). The Gamma function (Deacon et al. in Transp Res Rec J Transp Res Board 1806:9–18, 2002) is used to determine the properties from the repetitive shear tests. The CalME program is utilized to perform the incremental recursive simulations under three selected climate conditions, two levels of traffic loading and two pavement structures. On the basis of simulated data comparison was found the optimal correlation coefficient for UST device. Calculated p-value of Welch-modified two sample t-test for the distributions of predicted rutting from UST and SST test result suggests that surface rutting predicted by using UST and SST test results are statistically equal with high p values. Thus, the data from the proposed UST test equipment can be used for mechanistic-empirical (ME) pavement design and will provide rutting predictions that are statistically equal to the predictions for SST. The predicted rut depth was also found to be well correlated with accumulated equilibrium compliance.

Keywords: Shear properties · Permanent deformation · Rut · UST
Accumulated equilibrium compliance

1 Introduction

The increasing amount of heavy truck traffic together with increasing axle weights started to increase the demand for pavement structures with higher performance. Especially the heavy truck loading is a cause of pavement deterioration considering the

Z. Hossain et al. (eds.), *Solving Pavement and Construction Materials Problems with Innovative and Cutting-edge Technologies*, Sustainable Civil Infrastructures,
https://doi.org/10.1007/978-3-319-95792-0_2

pavement rutting. Flexible pavement permanent deformations in form of rutting reduce the quality of transportation and form safety risks. Surface rutting prevents the water from draining out of the pavement surface, standing water reduces the friction between pavement surface and the wheel, thus prolongs the breaking distance and enhances the risk of aquaplaning. Particularly on motorways and expressways, ruts can cause loss of driver's control over a vehicle at higher speeds during the overtaking. Thus the engineer's objective is to reduce permanent deformation by better predicting and monitoring the pavement rutting and taking the appropriate timely actions, such as reconstruction or rehabilitation, if needed.

Nowadays, an engineer has the option to choose from different test methodologies that suggest the link between the laboratory measured properties and field performance in regard of rutting (Von Quintus et al. 2012). The herein presented study is dealing with those test methods that has been used to develop transfer functions, and their effectiveness has been proven over the years of use by numerous researchers.

The permanent deformation response of asphalt mixtures was broadly studied as part of SHRP and reported in (Monismith et al. 1994). One of the research outcomes was the development of the Superpave Shear Tester (SST). The performance models, also referred as transfer functions, were further developed as part of WesTrack project (Monismith et al. 2000). Further the Superpave Shear Tester became a tool for mix design and The Repeated Shear Test at Constant Height (RSST-CH) is judged to be preferred method for analyzing the permanent deformations of asphalt mixtures (Harvey et al. 2002).

In the NCHRP 719 report, (Von Quintus et al. 2012), summarized transfer functions and recalibrated them using a broad database composed of full-scale test sections and results from Heavy Vehicle Simulators. The WesTrack transfer function was proved to be correlating well with asphalt mixtures' in situ rutting performance and implemented into MEPDG design program.

The Uniaxial Shear Tester was developed as part of the cooperation between University of California Pavement Research Center and Czech Technical University in Prague. The aim of the research was to develop a testing device that would allow to perform shear test in Universal Testing Machines (UTM), known also as Nottingham Asphalt Testers (NAT), so that the laboratories equipped with such a test device will be able to measure shear properties of asphalt mixtures (Zak et al. 2016). At the same time an innovative approach was proposed to calculate the accumulated equilibrium compliance as a linear viscoelastic parameter characterizing the long-life mixture rutting performance (Zak et al. 2013, 2018).

The aim of this study is to predict pavement rutting performance using CalME software (an ME analysis and design program for new flexible pavements and rehabilitation) and asphalt mixture shear properties measured by SST and UST test methods. The goal is to assess whether the properties measured by these test methods are able to simulate statistically equal rut depth predictions through the use of WesTrack transfer functions and a shear based incremental-recursive procedure.

It is a purpose of the article to present these correlations between two devices developed to measure shear properties of asphalt mixtures. Further the article proves that the data from the UST test equipment can be used for mechanistic-empirical (ME) pavement design and will provide rutting predictions that are statistically equal to the predictions for SST. The predicted rut depth was also found to be well correlated with accumulated equilibrium compliance.

2 Experiments

2.1 Material Specification

The simulations were performed with 3 asphalt mixtures. To make easier the orientation each asphalt mixture was denoted with number. Hot mix #1 contains 4.8% neat asphalt binder PG64-10 and 25% reclaimed asphalt, the nominal maximum aggregate size was 3/4″. The mixture was manufactured in the asphalt plant and small batches were taken from the construction site. From the mixture were prepared cylindrical specimens in Gyratory compactor, preferably was produced cylinder 135 mm high and 150 mm in diameter. The specimen was cut from the cylinder by double blade saw to obtain parallel sides. Further was the testing procedure in accordance with (AASHTO T 320-07 2011) in case of SST. The description of sample preparation for testing in UST can be found in (Zak et al. 2016). Mix #2 nominal maximum aggregate size is also 3/4″, it contains 15% of reclaimed asphalt and most importantly high polymer modified asphalt binder PG64-28. Batches of mix #2 were also taken from the construction site. Blocks of Mix #3 were cut from the UCPRC, Davis test section, from the blocks were drilled and cut the cylindrical samples. Then the samples were prepared as mentioned above. The mix #3 is 1/2 in. gap-graded rubberized hot-mix, it contains 7% asphalt binder PG64-10 without any reclaimed asphalt. The grading of aggregates is presented in Table 1.

2.2 Determination of Properties from Laboratory Measured Data

To obtain complete characteristics for rut depth predictions described in the following paragraphs were performed two tests. First the small amplitude oscillation tests were performed, known also as the frequency sweep shear tests. Data of the complex shear moduli at temperature at 3, 15, 30, 45 and 60 °C were selected and a data fit was performed to the measured frequencies in the range from 0.01 to 10 Hz. The dynamic moduli and their dependence on temperature and test frequency are described in form of sigmoidal function described by Eq. 1. Equation 1 defines the master curve as was presented in mechanistic-pavement design guide (MEPDG) (Hallin 2004) where T_r, reduced time of loading at the reference temperature is described as presented in Eq. 2 ("CalME Manual" 2011; Hallin 2004).

$$\log(E^*) = \delta + \gamma \frac{\alpha}{1 + \exp[\beta + \gamma * \log(T_r)]} \tag{1}$$

where E is the tensile modulus in MPa and $\alpha, \beta, \gamma,$ and δ are the determined constants. The tensile modulus is from measured complex shear modulus converted through the well-known relation from elastic theory, $E = 2G * (1 + v)$, where G is a measured shear modulus and v is a Poisson's ratio.

$$T_r = lt * \left(\frac{visc_{ref}}{visc}\right)^{aT} \tag{2}$$

Table 1. Asphalt mixture grading

Mixture type/sieve size	3/4″ (19 mm)	1/2″ (12.5 mm)	3/8″ (9.5 mm)	No. 4 (4.75 mm)	No. 8 (2.36 mm)	No. 16 (1.18 mm)	No. 30 (600 μm)	No. 50 (300 μm)	No. 100 (150 μm)	No. 200 (75 μm)	RAP
Mix#1 3/4″ HMA PG 64-10	100	78	63	45	32	27	19	15	10	5	25
Mix#2 3/4″ HMA PG64-28PM	100	88	76	48	34	27	20	14	8	5	15
Mix#3 1/2″ RHMA-G PG 64-10	100	98	83	40	23	–	12	–	–	5	–

Here lt is a loading time, $visc_{ref}$ is a binder viscosity at reference temperature, $visc$ is a binder viscosity at the present temperature and aT is a shift factor. Model parameters were determined in excel, when the Residual Sum of Squares (RSS) fit was performed using the solver. The measured properties were derived from five replicates.

Secondly the Repeated Shear Tests (RST) were performed. Prolong version of standard repeated shear test was performed when all the samples were loaded with 30,000 of repeated cycles. RST were run always on five replicates at 50 °C. Method developed by Deacon et al. (2002) was utilized to predict rutting of the asphalt layer. The laboratory data were fitted using the Gamma function:

$$y_p = A + \alpha * \left[1 - \exp\left(\frac{-\ln(N)}{\gamma}\right) * \left(1 + \frac{\ln(N)}{\gamma}\right)\right] * \exp\left(\frac{\beta + \tau}{\tau_{ref}}\right) \quad (3)$$

where N is a number of cycles, A, α, β, γ and δ are model parameters, τ is a shear stress, τ_{ref} is a reference shear stress. The final rut depth increment was calculated from the following equation ("CalME Manual" 2011; Deacon et al. 2002):

$$rd_i = K * h_i * y_p \quad (4)$$

Here rd_i is rut depth increment, K is a calibration factor and y_p is a plastic shear strain from Eq. 3. The models parameters were obtained performing the RSS in excel, when the optimum value was found with the help of solver. For the selected materials the measured properties reach its steady state trend in between first 100 to 300 cycles during Uniaxial Repeated Shear Test (URST) (Zak et al. 2016). As the test procedure is compound from 30,000 cycles, the performed RSS analysis tends to neglect the first loadings. In this regard a weight function was substituted into regression to give more weight in RSS to first set of 300 cycles. The weighting function (Eq. 5) with parameters a = 6 and b = −0.01 was used. The repeated shear tests were performed always with five replicates.

$$Weight = \exp[a + b * N * \ln(N)] \quad (5)$$

3 Incremental Recursive Simulation

3.1 Specification of Pavements and Environmental Conditions

The factorial of rut depths with 3 climate regions was run in CalME. The Inland Valley, Low Mountain and High Desert climate was chosen as a set expressing broad range of temperatures. Two traffic levels were selected. First denoted as TI10 where the pavement was loaded with 390,000 cycles in first year. The second traffic level was chosen to be TI 14 with 6,250,000 axles in pavement first year in service. The growth rate was selected to be 5% annually for both traffic levels. To bear such a loading were selected two pavement structures. The pavement structure was selected in the view of minimization of other distresses except the rutting. So the thick asphalt mixture layer on the thick unbound stiff aggregate base was selected. Selected pavement structures are in Fig. 1.

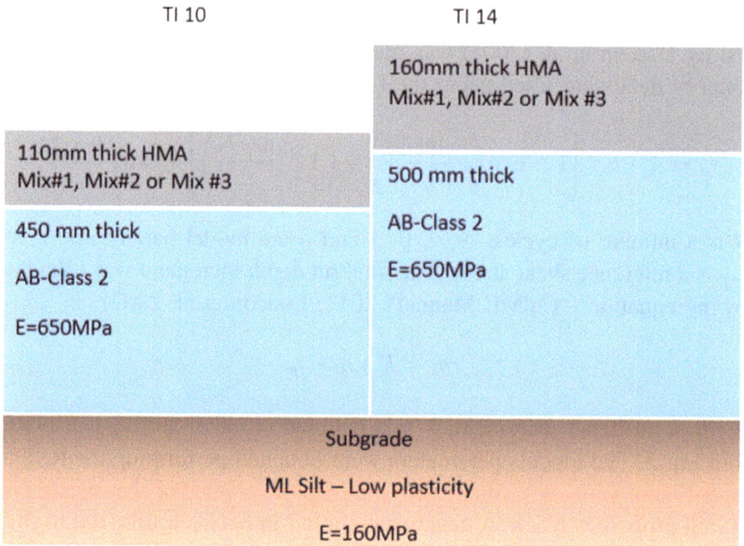

Fig. 1. Illustration of selected pavement structures

3.2 Incremental Recursive Simulation

In this approach, the pavement is assumed to behave as a multilayer elastic system. The idealization of a specific asphalt pavement permanent response was presented in Deacon et al. (2002). The principle is based on the calculation of rut depth estimate developed with traffic at depth of 50 mm below the outside edge of tire. The permanent shear strain in the asphalt layers is assumed to accumulate according to the principle relating the accumulated elastic shear strain measured in laboratory to permanent strain in pavement developed in Deacon et al. (2002). The computer program known as CalME has been used to perform the predictions (Ullidtz et al. 2010).

3.3 Uniaxial Shear Tester Calibration

The term K is a parameter that relates RD to plastic strain. The K parameter was first used during the WesTrack project (Monismith et al. 2000). Parameter K ranges from about 5.5 for a 150-mm layer to 10 for a 305-mm thick asphalt concrete layer (Deacon et al. 2002). Based on correlation between predicted RD was calculated parameter that relates the shear properties measured by UST to predicted plastic strain in pavements. The RSS was computed for each rut depth predictions at every condition and from these the average RSS was computed. Based on the RSS value minimization was found that the optimal K_{UST} parameter should be selected equal to 0.63 of its K equivalent ($K_{UST} = K * 0.63$).

The numerically predicted rutting occurred during first 14 days was more then 50% of total rutting after 40 years in some cases. The initial rutting depends highly on the *A* parameter of model presented in Eq. 3. However the A parameter can be selected widely if the other model parameters are changed accordingly without a significant effect on total RSS between model and experimentally measured data. In view of this the rutting predicted for first 14 days was omitted. As the reference point was used the first calculated point. This trimming neglected rutting in first 0.002% of total calculated pavement lifetime.

If the correlation coefficient K_{UST} was selected equal to 0.63 of its K equivalent and first 14 days of rutting was omitted then the predicted rutting is as presented in Figs. 2 and 3. The typical trend of calculated rut depths was selected for two climate conditions, High Desert and Low Mountain. As can be seen from both figures the predicted rutting correlates well.

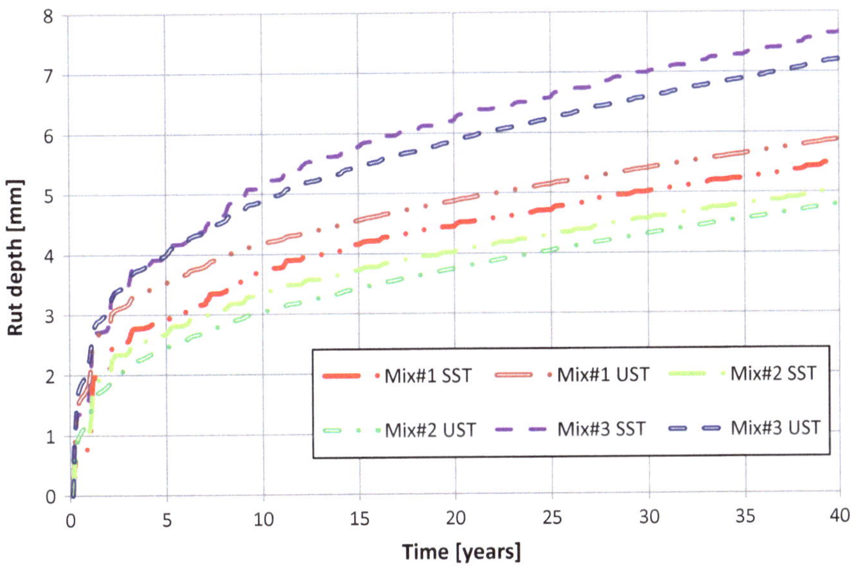

Fig. 2. Rut depth prediction, TI 14, High Desert climate

3.4 Welch-modified two sample t-test

In order to determine the statistical significance of the CalME predicted rut depth differences for UST and SST results, Welch-modified two sample t-test (Insightful 2001) is used. This statistical test is recommended to be used for evaluating small datasets (Ruxton 2006) and assumes unequal dataset variances. F1 and F2 are two distributions for predicted rut depths (from UST and SST test results) at 10th, 20th, and

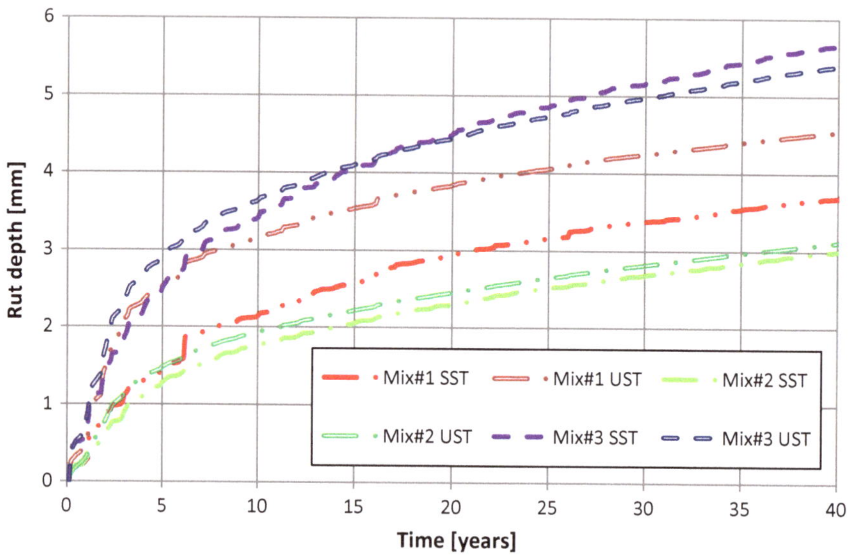

Fig. 3. Rut depth prediction, TI 10, Low Mountain climate

40th years for the climate regions and traffic levels of this study, the possible hypotheses and alternatives concerning these distributions are:

$$\begin{aligned} H0 &: F1(x) = F2(x) \\ HA &: F1(x) \neq \mathrm{F2(x)} \end{aligned} \tag{6}$$

Decision rule: Reject H0 if p-value < 0.10; Fail to reject H0 if p-value $\geq$ 0.10

Calculated p-value for the distributions of predicted rutting from UST and SST test results for years 10, 20, and 40 are 0.70, 0.95, and 0.91, respectively. This result suggests that surface rutting predicted by using UST and SST test results are statistically equal with high p-values. Thus, the data from the proposed UST test equipment can be used for mechanistic-empirical (ME) pavement design and will provide rutting predictions that are statistically equal to the predictions for SST.

Provided test results proves the correlation of incremental rutting simulation over 40 years. Which is more then the flexible pavement (wearing course especially) would last in real where the design period of the flexible pavements range from 20 to 30 years at best.

4 Correlation Between Accumulated Equilibrium Compliance and Predicted Rut Depth

The procedure describing the calculation of Accumulated Equilibrium Compliance (AEC), *Je, acc*, was proposed in (Zak et al. 2016). The accumulated creep compliance is calculated with the help of linear viscoelastic theory. The arbitrary rheological model can be used to calculate the retardation spectra. In view of the generally known principles that long-time processes are revealed in detail in the retardation spectrum and as the equilibrium compliance is obtained by integration over a discrete retardation spectrum it is suggest that such a material parameter express the long-life asphalt mixture performance regard to permanent deformation potential.

In this study was used 4-unit standard Kelvin-Voigt model, as was proposed in (Zak et al. 2018). The interval fitting procedure was utilized and the AECs were calculated for each asphalt mixture. The AEC values can be found in Table 2 in the appendix. If the linear regression between AEC and predicted rut depths is performed, then the correlation coefficient between these two variables may be expressed. The calculated correlation coefficients are listed in Table 3 in the appendix. The value represents correlation between six determined AECs from repeated shear tests in laboratory and six predicted rut depths at selected pavement during service period, loadings and climate conditions. The dependence on these conditions is presented in Fig. 4.

Table 2. Used material parameters

	Frequency sweep								
Designation	δ	β	γ	aT	A	VTS	Eref	tref	α
Mix#1_SST_314 HMA PG 64-10	1.699	1.139	0.783	0.719	9.631	−3.505	1316.472	30	2.483
Mix#2_SST_314 HMA PG 64-28PM	1.699	1.903	0.439	1.485	9.631	−3.505	479.100	30	3.937
Mix#3_SST_R-HMA-G	1.699	1.444	0.394	0.809	9.631	−3.505	961.776	30	3.937
Mix#1_UST_314 HMA PG 64-10	1.000	0.328	0.149	0.945	9.631	−3.505	351.863	30	3.182
Mix#2_UST_314 HMA PG 64-28PM	1.000	0.478	0.160	0.956	9.631	−3.505	278.045	30	3.182
Mix#3_UST_R-HMA-G	1.000	0.488	0.116	1.032	9.631	−3.505	235.455	30	3.182
–	Repeated shear						Accumulated equilibrium compliance, Je, acc [1/Pa]		
	A	α	tref	β	γ	δ			
Mix#1_SST_314 HMA PG 64-10	1.401	2.639	0.100	0.000	3.346	1.000	1.53E+11		
Mix#2_SST_314 HMA PG 64-28PM	1.458	1.749	0.100	0.000	2.920	1.000	2.39E+11		
Mix#3_SST_R-HMA-G	0.550	4.980	0.100	0.000	5.090	1.000	1.58E+11		
Mix#1_UST_314 HMA PG 64-10	0.525	3.390	0.100	0.000	1.504	1.000	1.75E+11		
Mix#2_UST_314 HMA PG 64-28PM	0.556	1.895	0.100	0.000	2.178	1.000	2.97E+11		
Mix#3_UST_R-HMA-G	0.125	3.772	0.100	0.000	2.780	1.000	1.20E+11		

Table 3. Calculated rut depths and correlation with accumulated equilibrium compliance

Conditions:	T14, Inland Valley						Rut depth and ACE correlation
Asphalt mixture and used test device	Mix#1 SST	Mix#1 UST	Mix#2 SST	Mix#2 UST	Mix#3 SST	Mix#3 UST	
Rutting after 10 years [mm]	4.54	4.53	3.85	3.23	5.86	5.22	−0.86
Rutting after 20 years [mm]	5.54	5.34	4.60	4.00	7.17	6.26	−0.84
Rutting after 40 years [mm]	6.70	6.37	5.64	5.08	8.74	7.62	−0.80
Conditions	T14, High Desert						
Rutting after 10 years [mm]	3.71	4.17	3.32	3.04	5.13	4.89	−0.82
Rutting after 20 years [mm]	4.48	4.88	4.05	3.76	6.26	5.89	−0.80
Rutting after 40 years [mm]	5.50	5.88	5.03	4.80	7.66	7.21	−0.77
Conditions	T14, Low Mountain						
Rutting after 10 years [mm]	3.05	3.58	3.05	2.97	4.32	4.44	−0.71
Rutting after 20 years [mm]	3.84	4.31	3.83	3.71	5.43	5.41	−0.70
Rutting after 40 years [mm]	4.89	5.34	4.94	4.77	6.78	6.68	−0.68
Conditions:	T10, Inland Valley						
Rutting after 10 years [mm]	2.27	3.08	1.52	1.60	3.19	3.38	−0.85
Rutting after 20 years [mm]	3.05	3.73	1.97	2.04	4.16	4.14	−0.87
Rutting after 40 years [mm]	3.80	4.39	2.49	2.58	5.23	5.02	−0.87
Conditions:	T10, High Desert						
Rutting after 10 years [mm]	3.18	4.12	2.50	2.27	4.12	4.16	−0.86
Rutting after 20 years [mm]	3.90	4.75	3.10	2.80	5.20	4.98	−0.87
Rutting after 40 years [mm]	4.74	5.51	3.82	3.51	6.44	5.97	−0.85
Conditions	T10, Low Mountain						
Rutting after 10 years [mm]	2.16	3.16	1.76	1.93	3.45	3.65	−0.76
Rutting after 20 years [mm]	2.93	3.84	2.30	2.45	4.51	4.45	−0.79
Rutting after 40 years [mm]	3.69	4.54	3.00	3.11	5.66	5.39	−0.78

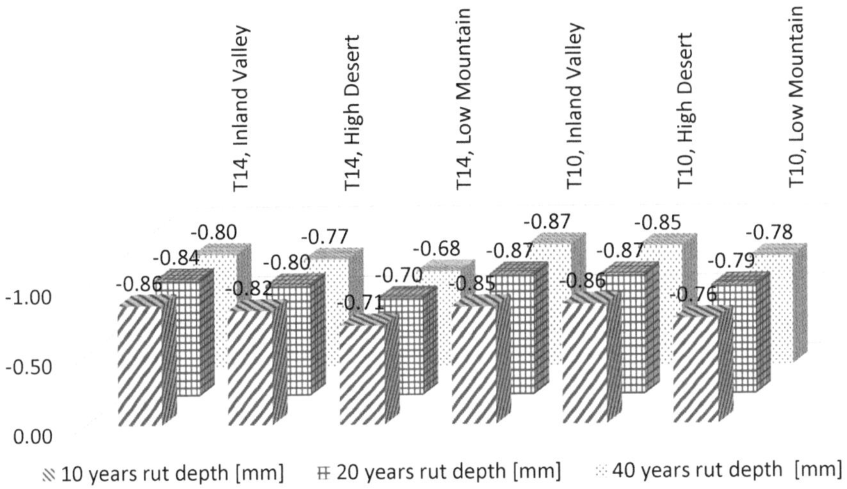

Fig. 4. Correlation between accumulated equilibrium compliance and simulated rut depth

The average correlation coefficient from the selected climate conditions, pavement structures and loadings after 10 years in service is −0.81. Consequently the correlation coefficient is equal to −0.81 after 20 years and −0.79 after 40 years in service. The negative correlation coefficient is evident, the higher AEC means less rutting potential. The overall average correlation coefficient is −0.8. Thus from the results of simulated rut depths and determined ACE may be concluded that the AEC is a suitable criterion for the asphalt mixture permanent deformation susceptibility.

5 Conclusions

It was purpose of this paper to compare the ability to determine the shear properties of newly developed UST device to currently used SST device through the pavement rutting predictions.

The incremental rutting predictions in CalME program were utilized for this purpose and the rut depth predictions were simulated for three asphalt mixtures in three climate conditions and with two selected traffic loading.

It can be concluded that the trends of predicted rut depths from properties determined from tests performed with UST and SST correlate well.

Calculated p-value for the distributions of predicted rutting from UST and SST test results for years 10, 20, and 40 suggests that surface rutting predicted by using UST and SST test results are statistically equal with high p-values. Consequently, the data from the developed UST test equipment can be used for mechanistic-empirical (ME) pavement design and will provide rutting predictions that are statistically equal to the predictions for SST.

It was found that the predicted rut depths correlate well also with the linear viscoelastic parameter, determined from repeated shear tests, the accumulated equilibrium compliance. Thus the computed accumulated equilibrium compliance is a suitable criterion for the assessment of the asphalt mixture susceptibility to permanent deformations.

Acknowledgements. The work reported here was supported by the project No. FV10526 in the program TRIO of the Ministry of Industry and Trade of the Czech Republic and also supported by both United States and Czech Republic governments through J. William Fulbright Commission.

References

AASHTO T 320-07: Standard Method of Test for Determining the Permanent Shear Strain and Stiffness of Asphalt Mixtures Using the Superpave Shear Tester (SST). American Association of State Highway and Transportation Officials (2011)

"CalME Manual". University of California, Berkeley (2011)

Deacon, J., Harvey, J., Guada, I, Popescu, L, Monismith, C.: Analytically based approach to rutting prediction. Transp. Res. Rec. J. Transp. Res. Board **1806** (1), 9–18 (2002). https://doi.org/10.3141/1806-02

Hallin, J.P.: NCHRP 1-37A Guide for Mechanistic-Empirical Design. National Cooperative Highway Research Program. http://apps.trb.org/cmsfeed/TRBNetProjectDisplay.asp?ProjectID=218 (2004)

Harvey, J., Guada, I., Monismith, C., Bejarano, M., Long, F.: Repeated Simple Shear Test for Mix Design: A Summary of Recent Field and Accelerated Pavement Test Experience in California. Copenhagen, Denmark (2002)

Insightful: S-Plus 6 for Windows User's Guide. Insightful Corporation, Seattle, Washington (2001)

Monismith, C.L., Hicks, R.G., Finn, F.N., Sousa, J., Harvey, J., Weissman, S., Deacon, J., Coplantz, J., Paulsen, G.: Permanent Deformation Response of Asphalt Aggregate Mixes, pp. 1–437 (1994 August)

Monismith, C.L, Deacon, J.A., Harvey, J.: WesTrack: Performance Models for Permanent Deformation and Fatigue. Pavement Research Center; Institute of Transportation Studies, University of California, Berkeley (2000)

Ruxton, G.D.: The unequal variance T-Test is an underused alternative to student's T-Test and the Mann-Whitney U test. Behav. Ecol. (2006). https://doi.org/10.1093/beheco/ark016

Ullidtz, P., Harvey, J.T., Basheer, I., Jones, D., Wu, R., Lea, J., Qing, Lu: CalME, a mechanistic-empirical program to analyze and design flexible pavement rehabilitation. Transp. Res. Rec. J. Transp. Res. Board **2010**(2153), 143–152 (2010). https://doi.org/10.3141/2153-16

Von Quintus, H., Jagannath, M., Bonaquist, R., Schwartz, Ch.W., Carvalho, R.L.: Calibration of Rutting Models for Structural and Mix Design. Transportation Research Board—National Research Council, Washington, D.C. (2012)

Zak, J., Stastna, J., Zanzotto, L., MacLeod, D.: Laboratory testing of paving mixes—dynamic material functions and wheel tracking tests. Int. J. Pavement Res. Technol. **3**(6), 147–154 (2013). https://doi.org/10.6135/ijprt.org.tw/2013.6(3).147

Zak, J., Monismith, C.L., Coleri, E., Harvey, J.T.: Uniaxial shear tester—test method to determine shear properties of asphalt mixtures. Road Mater. Pavement Des. J., 87–103 (2016). https://doi.org/10.1080/14680629.2016.1266747

Zak, J, Coleri, E., Stastna, J., Harvey, J.T.: Accumulated equilibrium compliance as a rutting susceptibility parameter. Submitted to Road Mater. Pavement Des. J. (2018)

Thermal Gradient in Self Compacting Concrete—An Experimental Investigation

Abhijeet S. Gandage[1(✉)] and V. Vinayaka Ram[2]

[1] Department of Civil Engineering, MAEER's Maharashtra Institute of Technology, Pune, India
abhijeetgandage@hotmail.com

[2] Department of Civil Engineering, BITS Pilani Hyderabad Campus, Hyderabad, India

Abstract. The temperature differential between top and bottom of concrete pavement develops curling stresses in the pavement slab. These stresses are tensile in nature and together with wheel loads influence the performance of concrete pavements. The thermal performance of concrete is influenced by the thermal properties of cement paste and aggregates. Self Compacting Concrete (SCC) is a widely used High Performance Concrete (HPC) in major infrastructure projects. It is characterized by high powder content, reduced coarse aggregate content, lower water cement ratio and use of High Range Water Reducing (HRWR) admixtures. An experimental study was undertaken to investigate the development of thermal gradient in SCC using manufactured sand with pozzolan addition (flyash as cement replacement) and perlite (insulating material as fine aggregate replacement). This paper present the findings of the study undertaken.

Keywords: Thermal gradient · Self compacting concrete · Flyash
Perlite

1 Introduction

Cement concrete is a widely used construction material in transportation infrastructure development. It is used in pavement construction since 1890 (Delatte 2008). Concrete pavements designed and constructed prior to 1930s were mainly analyzed for traffic loads. The only environment related factor that affected concrete pavements was freeze-thaw effect. This problem was addressed by air entrained concrete. The temperature related loads were not considered in the analysis. Westergaard (1927) first proposed linear profile of temperature differential between top and bottom for an infinite concrete slab. The temperature differential developed leads to curling stress. Bradbury attempted to improve Westergaard's solution for temperature stresses by suggesting suitable coefficients for a finite concrete slab. Teller and Sutherland, 1935, (Fwa 2006) reported the non-linear nature of temperature distribution developed in the concrete pavement. It was proposed that the stresses due to temperature loads are equally important as that of wheel loads. The temperature differential gradient developed and vehicle loads influence the service life of the concrete pavement (Kook et al.).

Z. Hossain et al. (eds.), *Solving Pavement and Construction Materials Problems with Innovative and Cutting-edge Technologies*, Sustainable Civil Infrastructures,
https://doi.org/10.1007/978-3-319-95792-0_3

The temperature differential developed in concrete slab depends on air temperature, solar radiation, humidity and moisture (Rao and Roesler 2005).

2 Research Significance

Concrete pavements offer long lasting surface with high load carrying capacity and low maintenance. The vehicle wheel loads and environmental loads (temperature and moisture variations) influence the performance and maintenance regime of the concrete pavements (Chai and van Staden 2012). Curling stress in cement concrete pavement occur on account of temperature differential. The self weight of the concrete slab and superimposed loads restrain the slab movement, leading to curling stress in the slab. Curling stress in concrete pavements influences crack development, roughness and long term serviceability problems. The inclusion of curling stress in the concrete pavement analysis is crucial for zero maintenance concrete pavements (Huang 2004). The studies undertaken so far focused on normal concrete. An attempt has been made to study the development of thermal gradient in case of SCC with manufactured sand.

3 Self Compacting Concrete (SCC)

SCC is a type of HPC that flows under its own weight with adequate resistance to segregation. The segregation resistance of this concrete is sensitive to the mix design and powder composition of the mix (Goodier 2003). SCC is similar to normal concrete with respect to the materials of production, with the difference being that SCC has ability to flow as compared to normal concrete. SCC finds its origin in late 1980s in Japan for the underwater concrete applications. Due to its numerous operational advantages over traditionally vibrated concrete, it is now a preferred concrete option in major infrastructure projects (Domone 2006). SCC mix design was first conceptualized by Okamura and Ozawa (Gaimster and Dixon 2003). The minimum guidelines for SCC mix design were first published in 2005 (EFNARC 2005). High cement content, reduced coarse aggregate content, increased fines and use of HRWR superplasticizers are the main distinguishing factors of SCC with respect to normally vibrated concrete. With the use of HRWR superplasticizers, water cement ratio as low as 0.3 can be adopted in SCC mix design. The mix design process of SCC can be admixture based, powder based or combination of admixture and powder methods (Türkel and Ali 2010). In the present study, powder based mix design of SCC has been adopted.

The thermal properties of concrete influencing its performance include coefficient of thermal expansion, specific heat of concrete and thermal conductivity. The thermal properties of concrete depend on the thermal properties of individual material used in the concrete mix. Due to high cement content in the concrete mix, the thermal conductivity will increase (Demirboğa and Gul 2003). The powder additions, like flyash, used as cement replacement material, reduce the thermal conductivity due to the insulating nature of the powder additions.

The parent rock of coarse aggregates and fine aggregates (manufactured sand) used in the present study is granite. The mineral composition of this rock is primarily crystalline. Hence, the thermal conductivity of concrete using granite based aggregates is more. Aggregates with crystalline siliceous mineral composition have more thermal conductivity values than aggregates with amorphous or carbonate based mineral structure (Kodur and Sultan 2003).

4 Experimental Program

The experimental program comprised of two main parts viz.

1. Mix design of M-40 grade SCC with various additions.
2. Thermal gradient measurement.

4.1 Mix Design of M-40 Grade SCC

In the present study, M-40 grade SCC with manufactured sand is adopted. Class C flyash is used as a cement replacement additive. From the experimental trials undertaken prior to this study, the appropriate cement replacement dosage for the study mix adopted was 20% (Gandage et al. 2013a, b).

To understand the effect of use of light weight insulating aggregate replacement material, perlite was adopted as a replacement of fine aggregates. Based on the experiments undertaken, the appropriate dosage of perlite as a fine aggregate replacement material considered is 5% (Gandage et al. 2013a, b).

The SCC mix design was undertaken as per guidelines mentioned in EFNARC. Polycarboxylate ether based HRWR superplasticizer was used in the concrete mix. The fresh properties viz. slump flow and T_{50} time were measured during the preparation stage (Fig. 1). As per IS 10262: 2009 guidelines, the 28-day target compressive strength for M-40 grade concrete is 48.25 MPa. As per IRC 58: 2011 guidelines, the minimum value for 28 day flexural strength for concrete to be used in pavement applications should be 4.5 MPa.

Fig. 1. Slump flow test for SCC mix

Table 1 specifies the mix design adopted, fresh state properties and mechanical properties (28 days) achieved for the three specimens.

Table 1. Mix design, fresh state properties and mechanical properties for M-40 grade SCC with various additions

Material	Mix A	Mix B	Mix C
Cement (kg/m^3)	430	344	344
Flyash (kg/m^3)	0	86	86
Flyash dosage (% cement)	0	20	20
F. A. (kg/m^3)	1135	1135	1078.25
Perlite (kg/m^3)	0	0	56.75
Perlite dosage (% FA replacement)	0	0	5
C. A. (kg/m^3)	630	630	630
Water (kg/m^3)	200	200	200
HRWR (l/100 kg of binder)	0.8	0.8	0.8
Slump flow (mm)	587	592	597
T_{50} (s)	6	8	8
28-day compressive strength (MPa)	57.78	50.82	50.67
28-day flexural strength (MPa)	7.34	6.54	8.40

Mix A represents M-40 grade SCC without any additions. Mix B represents M-40 grade SCC with 20% flyash as a cement replacement material. Mix C represents M-40 grade SCC with 20% flyash as a cement replacement material and 5% perlite as a fine aggregate replacement material. All the three mixes are prepared using manufactured sand.

4.2 Thermal Gradient Measurement

The thermal gradient developed in concrete was measured by preparing three cube specimens of 150 mm dimension made of the above mentioned three mixes. The cubes were cured for 28 days and subjected to environmental conditions as shown in Fig. 2.

The thermal gradient developed in the cubes was measured using resistance temperature detector (RTD) type sensors. Five sensors of lengths measuring 2.5, 5, 7.5, 10 and 12.5 cm were placed in fresh concrete (Fig. 3). For each cube in all 15 sensors were embedded (3 numbers of each length). The layout of the sensors in the cube is shown in Fig. 3. In addition to these sensors, surface sensor was attached on the surface of the cube to measure the surface temperature. All the sensors were attached to a 16-channel data logger, programmed to scan and save the temperature data at an interval of 60 s. Temperature data for a period of 48 h was scanned and saved in the on-board memory of the data logger. At the end of the 48 h period, the data was retrieved and data loggers were again operational for a fresh cycle of 48 h.

A total of 75 day observations were recorded. Out of the total number of observations recorded, representative three typical cases, as tabulated in Table 2, are discussed in the paper.

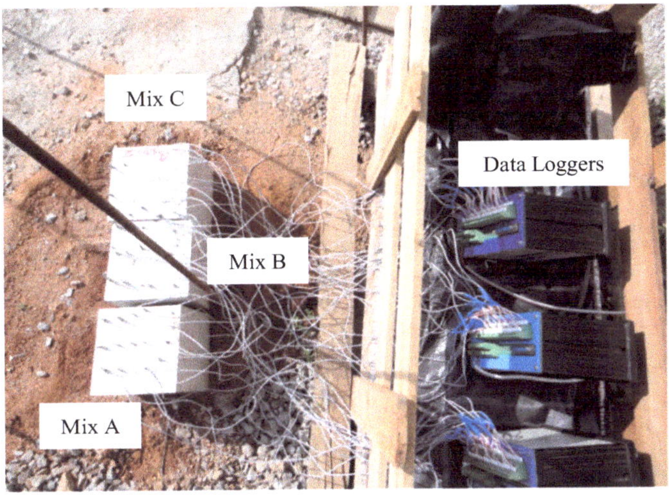

Fig. 2. Cube specimens with embedded sensors connected to data loggers

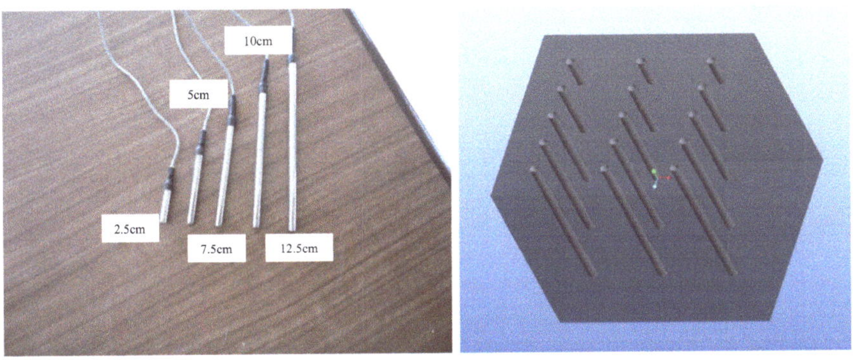

Fig. 3. RTD sensors of varying lengths and its layout in SCC cube specimen

Table 2. Details of the Cases

Case No.	Case particulars	Date
1	Maximum ambient temperature	30–31 October 2013
2	Minimum ambient temperature	13–14 November 2013
3	24 h precipitation	25–26 October 2013

Tables 3, 4 and 5 represents the average temperature values recorded for a period of 24 h for each case and all three mixes.

Figures 4, 5, and 6 represents the graphical plots for temperature profile for each of the cases and all three mixes.

Table 3. 24 h temperature recording for Case 1 (max. temperature) in Mix A, B and C

Sr. No.	Mix A						Mix B					
	Temperature values (°C)						Temperature values (°C)					
	Surface	2.5 cm	5 cm	7.5 cm	10 cm	12.5 cm	Surface	2.5 cm	5 cm	7.5 cm	10 cm	12.5 cm
1	35.58	32.70	30.62	29.55	28.70	28.74	37.67	32.68	31.21	29.33	28.39	27.57
2	39.32	36.23	34.35	33.28	32.33	32.14	40.02	35.16	33.95	32.13	31.07	30.18
3	40.40	38.00	36.60	35.84	35.00	34.70	39.51	35.78	35.11	33.72	32.82	32.04
4	41.55	39.19	37.84	37.11	36.32	35.99	40.37	36.68	35.93	34.62	33.81	33.12
5	40.28	39.15	38.28	37.87	37.12	36.78	36.53	35.51	35.60	34.87	34.29	33.78
6	37.02	36.82	36.73	36.85	36.47	36.46	33.25	33.29	33.80	33.71	33.46	33.32
7	33.54	34.25	34.71	35.22	35.06	35.25	30.08	30.90	31.78	32.15	32.17	32.31
8	30.67	31.61	32.37	33.12	33.15	33.50	27.75	28.79	29.85	30.51	30.73	31.09
9	28.42	29.37	30.17	30.99	31.11	31.56	26.04	27.10	28.21	28.99	29.31	29.78
10	26.70	27.62	28.38	29.19	29.35	29.87	24.65	25.73	26.80	27.62	28.00	28.52
11	25.25	26.20	26.88	27.69	27.86	28.42	23.58	24.59	25.59	26.42	26.81	27.34
12	24.24	25.07	25.67	26.49	26.65	27.21	22.89	23.81	24.71	25.44	25.85	26.41
13	23.33	24.16	24.73	25.45	25.63	26.26	22.20	23.14	23.95	24.67	25.06	25.59
14	22.55	23.33	23.85	24.64	24.81	25.37	21.76	22.70	23.41	24.04	24.47	24.99
15	22.00	22.77	23.22	23.91	24.08	24.71	21.49	22.25	22.98	23.54	23.92	24.46
16	21.61	22.32	22.73	23.39	23.52	24.15	21.08	21.84	22.51	23.09	23.43	23.91
17	21.18	21.83	22.24	22.96	23.10	23.67	20.61	21.39	22.02	22.62	23.00	23.44
18	20.69	21.43	21.78	22.47	22.63	23.22	20.15	20.94	21.57	22.13	22.51	23.02
19	20.26	20.98	21.39	22.00	22.14	22.78	19.91	20.55	21.19	21.71	22.05	22.56
20	19.97	20.61	20.99	21.63	21.74	22.37	19.71	20.33	20.90	21.39	21.72	22.20
21	19.94	20.44	20.73	21.37	21.49	22.03	20.06	20.34	20.80	21.24	21.57	21.95
22	21.25	21.05	21.01	21.45	21.48	21.99	22.54	21.50	21.49	21.54	21.72	22.00

(*continued*)

Table 3. *(continued)*

Sr. No.	Mix A						Mix B					
	Temperature values (°C)						Temperature values (°C)					
	Surface	2.5 cm	5 cm	7.5 cm	10 cm	12.5 cm	Surface	2.5 cm	5 cm	7.5 cm	10 cm	12.5 cm
23	24.39	23.79	23.02	22.97	22.71	23.16	28.39	24.41	23.60	22.97	22.84	22.87
24	27.74	26.46	25.43	25.17	24.77	25.16	29.26	26.78	25.72	24.58	24.16	23.87

Sr. No.	Mix C						
	Temperature values (°C)						
	12.5 cm	Surface	2.5 cm	5 cm	7.5 cm	10 cm	12.5 cm
1	27.57	35.44	31.39	29.68	28.04	26.67	25.96
2	30.18	37.83	34.13	32.40	30.67	29.02	28.11
3	32.04	37.89	35.38	34.03	32.56	30.93	29.97
4	33.12	38.65	36.25	34.90	33.59	32.07	31.20
5	33.78	36.80	35.84	35.16	34.28	32.97	32.19
6	33.32	33.65	33.62	33.59	33.36	32.67	32.18
7	32.31	30.76	31.20	31.70	31.99	31.75	31.53
8	31.09	28.42	28.92	29.74	30.43	30.61	30.70
9	29.78	26.65	27.13	28.07	28.94	29.41	29.68
10	28.52	25.17	25.63	26.58	27.48	28.11	28.49
11	27.34	23.95	24.39	25.29	26.21	26.90	27.31
12	26.41	23.16	23.52	24.35	25.17	25.92	26.40
13	25.59	22.45	22.81	23.56	24.37	25.10	25.55
14	24.99	21.93	22.28	23.03	23.73	24.52	24.97
15	24.46	21.58	21.86	22.58	23.27	23.99	24.47
16	23.91	21.22	21.49	22.11	22.82	23.49	23.93
17	23.44	20.74	21.06	21.69	22.34	23.06	23.47

(continued)

Table 3. *(continued)*

Sr. No.	Mix C						
	Temperature values (^{o}C)						
	12.5 cm	Surface	2.5 cm	5 cm	7.5 cm	10 cm	12.5 cm
18	23.02	20.30	20.60	21.27	21.87	22.62	23.07
19	22.56	19.98	20.20	20.85	21.50	22.16	22.63
20	22.20	19.70	19.90	20.54	21.15	21.78	22.23
21	21.95	19.38	19.79	20.38	20.92	21.58	22.10
22	22.00	21.44	20.69	20.82	21.19	21.70	21.87
23	22.87	25.96	23.65	22.90	22.39	22.31	22.38
24	23.87	28.43	26.53	25.57	24.69	24.08	23.80

Table 4. 24 h temperature recording for Case 2 (min. temperature) in Mix A, B and C

Sr. No.	Mix A						Mix B					
	Temperature values (°C)						Temperature values (°C)					
	Surface	2.5 cm	5 cm	7.5 cm	10 cm	12.5 cm	Surface	2.5 cm	5 cm	7.5 cm	10 cm	12.5 cm
1	28.49	27.35	25.87	25.18	24.53	24.62	35.88	30.38	28.73	27.15	26.49	25.96
2	33.39	30.25	28.85	28.16	27.46	27.47	37.65	32.18	31.17	29.73	28.97	28.34
3	34.97	32.50	31.11	30.48	29.77	29.58	37.85	33.37	32.67	31.43	30.71	30.10
4	37.57	34.76	33.35	32.65	31.84	31.53	36.29	33.55	33.22	32.31	31.77	31.37
5	37.98	36.07	34.82	34.24	33.42	33.08	34.03	32.95	32.96	32.43	32.10	31.81
6	36.48	35.55	34.89	34.66	34.09	33.74	30.16	30.96	31.52	31.55	31.43	31.38
7	33.23	33.75	33.82	34.00	33.63	33.45	26.44	28.36	29.39	29.88	30.03	30.20
8	29.22	30.68	31.44	32.05	31.99	32.04	23.82	25.83	27.05	27.81	28.15	28.48
9	25.82	27.40	28.45	29.37	29.49	29.76	21.93	23.88	25.08	25.93	26.36	26.79
10	23.45	24.90	25.93	26.90	27.11	27.57	20.63	22.45	23.54	24.38	24.84	25.31
11	21.70	23.05	23.97	24.93	25.16	25.71	19.53	21.27	22.27	23.11	23.57	24.10
12	20.40	21.62	22.46	23.38	23.61	24.25	18.61	20.30	21.26	22.05	22.54	23.09
13	19.39	20.49	21.29	22.15	22.40	23.07	17.89	19.53	20.39	21.20	21.67	22.22
14	18.49	19.64	20.32	21.20	21.43	22.10	17.30	18.83	19.70	20.43	20.94	21.50
15	17.87	18.87	19.57	20.36	20.60	21.32	16.77	18.27	19.07	19.82	20.25	20.84
16	17.36	18.33	18.93	19.75	19.96	20.64	16.22	17.70	18.47	19.22	19.72	20.23
17	16.81	17.84	18.39	19.18	19.42	20.10	15.80	17.23	17.99	18.65	19.15	19.74
18	16.36	17.27	17.89	18.62	18.85	19.59	15.22	16.76	17.51	18.22	18.66	19.26
19	15.85	16.79	17.40	18.18	18.37	19.09	14.96	16.38	17.07	17.79	18.27	18.81
20	15.40	16.38	16.91	17.71	17.95	18.64	14.90	16.17	16.76	17.41	17.90	18.39
21	15.13	16.06	16.59	17.30	17.53	18.23	16.81	16.67	16.87	17.30	17.72	18.25
22	15.56	16.12	16.48	17.07	17.24	18.02	23.97	20.18	19.28	18.78	18.83	18.99

(*continued*)

Table 4. *(continued)*

Sr. No.	Mix A						Mix B					
	Temperature values (°C)						Temperature values (°C)					
	Surface	2.5 cm	5 cm	7.5 cm	10 cm	12.5 cm	Surface	2.5 cm	5 cm	7.5 cm	10 cm	12.5 cm
23	20.44	18.88	18.07	18.16	18.10	18.85	29.07	24.44	23.12	21.95	21.58	21.35
24	26.39	23.96	22.37	21.84	21.38	21.87	33.35	27.85	26.32	24.80	24.20	23.75

Sr. No.	Mix C					
	Temperature values (°C)					
	Surface	2.5 cm	5 cm	7.5 cm	10 cm	12.5 cm
1	31.66	28.65	26.99	25.58	24.55	24.12
2	34.72	31.30	29.80	28.33	26.95	26.27
3	35.35	32.61	31.40	30.17	28.81	28.11
4	35.04	33.08	32.20	31.21	30.00	29.35
5	34.29	32.98	32.38	31.76	30.79	30.21
6	31.93	31.69	31.60	31.40	30.77	30.33
7	28.94	29.41	29.93	30.27	30.07	29.91
8	26.12	26.91	27.89	28.64	28.86	28.92
9	24.04	24.77	25.87	26.82	27.33	27.58
10	22.30	23.03	24.10	25.15	25.79	26.18
11	21.03	21.69	22.82	23.72	24.59	25.03
12	19.98	20.57	21.58	22.57	23.40	23.92
13	19.10	19.68	20.64	21.58	22.47	23.03
14	18.34	18.90	19.84	20.75	21.65	22.22
15	17.74	18.25	19.14	20.04	20.96	21.54
16	17.11	17.65	18.49	19.38	20.28	20.90
17	16.58	17.04	17.93	18.77	19.74	20.31

(continued)

Table 4. *(continued)*

Sr. No.	Mix C					
	Temperature values (°C)					
	Surface	2.5 cm	5 cm	7.5 cm	10 cm	12.5 cm
18	16.13	16.61	17.44	18.30	19.23	19.85
19	15.66	16.20	17.01	17.88	18.78	19.43
20	15.35	15.84	16.63	17.45	18.39	18.98
21	15.43	15.74	16.42	17.13	18.06	18.61
22	18.55	17.41	17.33	17.49	18.11	18.52
23	24.28	21.60	20.65	19.83	19.59	19.56
24	28.48	25.32	23.90	22.69	21.88	21.51

Table 5. 24 h temperature recording for Case 3 (24 h precipitation) in Mix A, B and C

Sr. No.	Mix A						Mix B					
	Temperature values (°C)						Temperature values (°C)					
	Surface	2.5 cm	5 cm	7.5 cm	10 cm	12.5 cm	Surface	2.5 cm	5 cm	7.5 cm	10 cm	12.5 cm
1	26.40	25.69	25.22	25.24	25.06	25.30	22.27	22.10	22.28	22.35	22.47	22.63
2	27.01	26.29	25.88	25.94	25.75	25.99	22.38	22.12	22.29	22.34	22.44	22.60
3	26.99	26.54	26.22	26.32	26.14	26.36	22.72	22.37	22.51	22.51	22.57	22.73
4	26.90	26.63	26.31	26.43	26.27	26.54	22.98	22.64	22.77	22.74	22.79	22.92
5	26.70	26.62	26.47	26.67	26.54	26.77	22.97	22.69	22.90	22.91	22.95	23.08
6	25.08	25.68	25.84	26.25	26.23	26.58	22.62	22.43	22.72	22.79	22.89	23.04
7	22.84	23.46	24.13	24.86	25.05	25.50	22.32	22.15	22.43	22.56	22.68	22.89
8	22.25	22.63	23.06	23.67	23.87	24.49	22.13	21.97	22.25	22.40	22.54	22.76
9	22.20	22.46	22.73	23.27	23.39	23.96	22.00	21.85	22.12	22.27	22.41	22.63
10	22.13	22.36	22.55	23.07	23.17	23.64	21.91	21.80	22.02	22.15	22.29	22.53
11	22.02	22.26	22.42	22.92	23.01	23.48	21.81	21.75	21.95	22.08	22.22	22.45
12	22.08	22.27	22.39	22.84	22.93	23.40	21.79	21.66	21.88	22.00	22.12	22.35
13	22.10	22.27	22.38	22.83	22.90	23.34	21.79	21.63	21.83	21.93	22.06	22.28
14	22.08	22.27	22.36	22.82	22.86	23.29	21.76	21.62	21.83	21.89	22.01	22.23
15	22.03	22.22	22.29	22.68	22.74	23.11	21.70	21.55	21.77	21.84	21.95	22.16
16	22.23	22.38	22.36	22.74	22.75	23.15	21.63	21.52	21.69	21.75	21.86	22.03
17	22.30	22.43	22.46	22.83	22.83	23.23	21.60	21.49	21.62	21.69	21.78	21.97
18	22.22	22.39	22.43	22.83	22.83	23.26	21.60	21.41	21.54	21.63	21.73	21.88
19	22.07	22.29	22.35	22.76	22.78	23.16	21.48	21.32	21.47	21.54	21.63	21.79
20	21.86	22.07	22.16	22.55	22.55	22.94	21.34	21.21	21.41	21.46	21.58	21.74
21	21.91	22.05	22.08	22.44	22.44	22.87	21.45	21.27	21.42	21.44	21.58	21.73
22	22.25	22.32	22.26	22.58	22.55	22.94	21.73	21.52	21.59	21.60	21.67	21.81

(*continued*)

Table 5. *(continued)*

Sr. No.	Mix A						Mix B					
	Temperature values (°C)						Temperature values (°C)					
	Surface	2.5 cm	5 cm	7.5 cm	10 cm	12.5 cm	Surface	2.5 cm	5 cm	7.5 cm	10 cm	12.5 cm
23	22.62	22.67	22.56	22.84	22.76	23.12	22.28	21.82	21.88	21.81	21.85	21.99
24	22.90	22.90	22.77	23.03	22.92	23.26	23.57	22.62	22.55	22.34	22.32	22.40

Sr. No.	Mix C					
	Temperature values (°C)					
	Surface	2.5 cm	5 cm	7.5 cm	10 cm	12.5 cm
1	22.55	22.38	22.37	22.52	22.61	22.93
2	22.80	22.45	22.53	22.57	22.63	22.72
3	22.94	22.68	22.74	22.80	22.84	22.90
4	22.90	22.71	22.82	22.92	22.92	23.03
5	22.52	22.42	22.62	22.79	22.88	22.97
6	22.22	22.12	22.35	22.56	22.75	22.88
7	21.98	22.00	22.16	22.38	22.60	22.76
8	21.91	21.82	22.03	22.25	22.48	22.62
9	21.86	21.73	21.93	22.13	22.39	22.54
10	21.71	21.71	21.87	22.08	22.32	22.45
11	21.70	21.68	21.80	22.00	22.22	22.38
12	21.67	21.72	21.77	21.95	22.18	22.31
13	21.67	21.63	21.77	21.90	22.14	22.25
14	21.60	21.59	21.73	21.88	22.06	22.21
15	21.60	21.50	21.65	21.79	21.99	22.10
16	21.60	21.43	21.59	21.72	21.92	22.03
17	21.56	21.40	21.54	21.63	21.86	21.94

(continued)

Table 5. *(continued)*

Sr. No.	Mix C					
	Temperature values (^oC)					
	Surface	2.5 cm	5 cm	7.5 cm	10 cm	12.5 cm
18	21.42	21.29	21.48	21.60	21.77	21.85
19	21.29	21.20	21.37	21.53	21.70	21.77
20	21.45	21.28	21.42	21.55	21.71	21.78
21	21.75	21.54	21.60	21.65	21.81	21.88
22	22.38	21.87	21.87	21.89	21.98	22.04
23	23.60	22.78	22.61	22.46	22.42	22.41
24	23.83	23.16	23.11	23.00	22.87	22.83

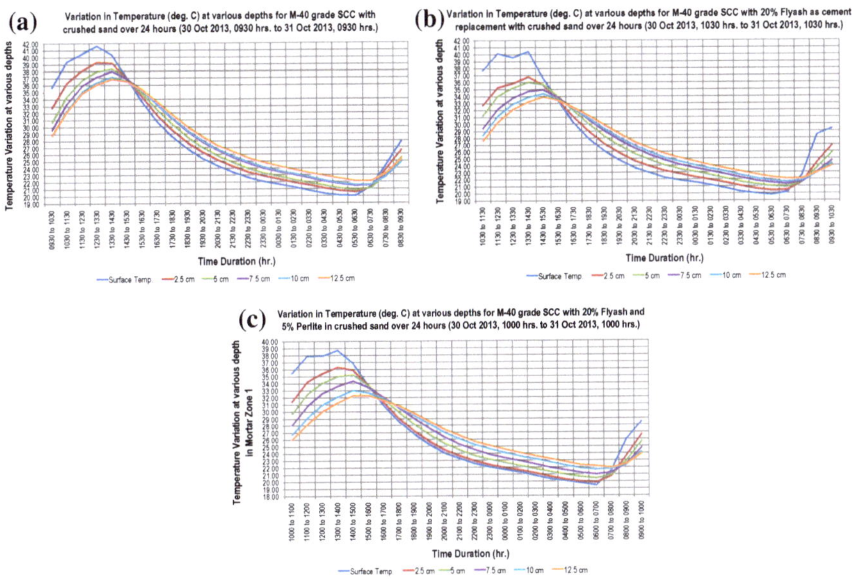

Fig. 4. Temperature plots for a period of 24 h for Case 1 in Mix A, B and C

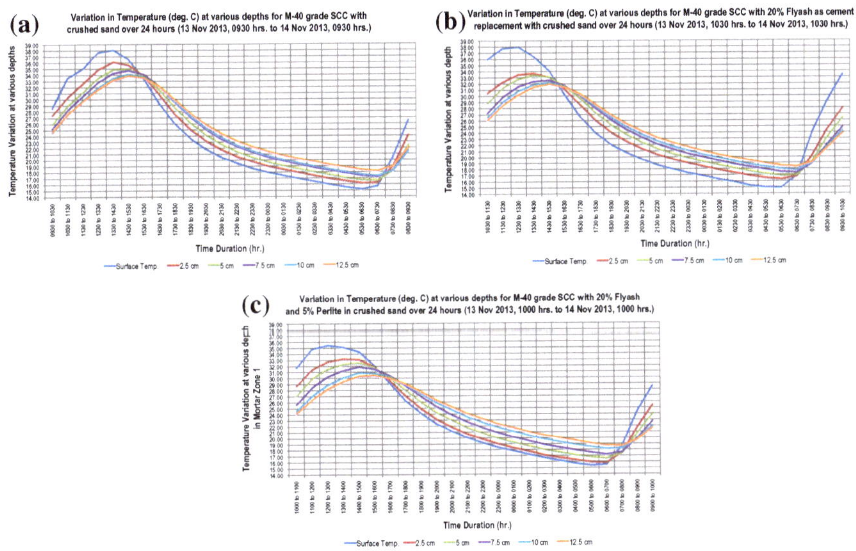

Fig. 5. Temperature plots for a period of 24 h for Case 2 in Mix A, B and C

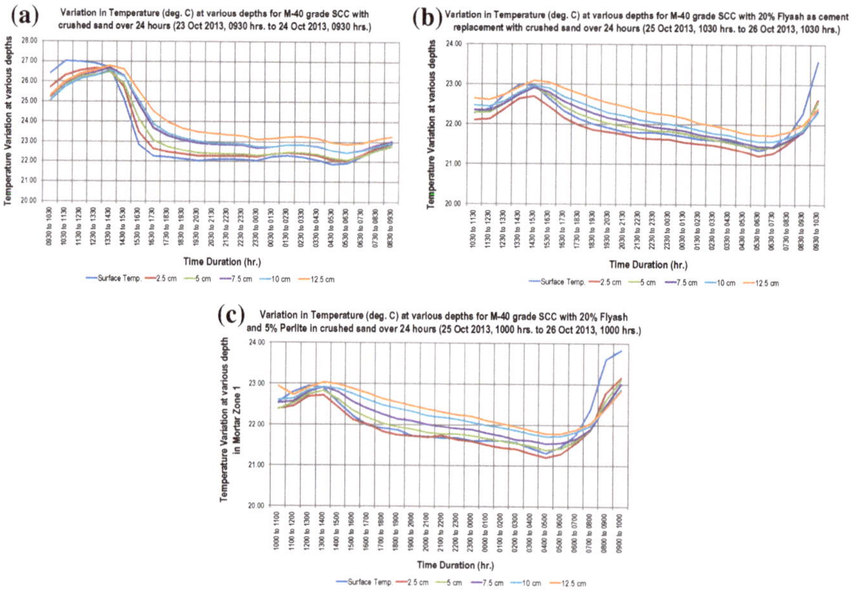

Fig. 6. Temperature plots for a period of 24 h for Case 3 in Mix A, B and C

5 Results and Discussion

Curling stress developed in cement concrete pavements depend on the temperature gradient developed between the top and bottom of the pavement slab. In the present study, a cube of 15 cm side is considered and the temperature difference developed under three different cases, as mentioned in the preceding section, is discussed.

For Case 1, a peak surface temperature recorded for Mix A, B and C were 41.55, 40.37 and 38.65 °C respectively. The corresponding ambient air temperature was 32 °C. The minimum surface temperature recorded on the same day for all the three cubes were 19.94, 19.71 and 19.79 °C respectively. The corresponding ambient air temperature recorded was 19 °C.

Case 2 represents the minimum temperature recorded. The minimum surface temperatures recorded were 15.13, 14.9 and 15.35 °C for Mix A, B and C respectively. The maximum surface temperatures recorded for the particular day were 37.98, 37.85 and 35.35 °C for Mix A, B and C respectively. The maximum ambient air temperature recorded was 29 °C and minimum ambient air temperature was 17 °C.

In Case 3, it is observed that the temperature profile is more or less parallel to abscissa for all the three mixes. Due to 24 h precipitation, overall ambient temperatures dropped down. Also as the cubes were exposed to the precipitation, the drop in temperature in the cube interior is also observed. For Mix A, the drop in temperature is steep, from an average peak of 26.5 °C at 1430 h to an average of 22.5 °C at 1630 h, spread over a period of 2 h. Thereafter, the temperatures remain in the range of 22–23 °C for the remainder of the time period. In case of Mix B and C, the drop in temperature is not significant. The average temperatures fluctuate between 23 °C (at 1400 h) to a minimum of 21.5 °C (at 0500 h). It is observed that in case of Mix B and C, the drop in temperature is gradual and spread over a period of 16 h. It is further observed that after 24 h, when the precipitation intensity decreased, the temperatures begin to rise, particularly for Mix B and C.

The temperature difference was measured between the surface of the cube and at a point 12.5 cm in the cube. The results of temperature difference (ΔT, °C) and the gradient (ΔT/d, °C/mm) are tabulated as in Table 6.

It is observed from the above result table, that the temperature difference (day as well as night time) developed in Mix A for Case 1 is low as compared to Mix B and C. This is accounted for the insulating property of flyash and perlite being added in Mix B and C respectively. The corresponding gradients developed are less as compared to the values reported in the Arlington Road Test (Huang 2004). For a 152 mm slab, the maximum temperature differential reported in the test was 12.2 °C and the corresponding temperature gradient developed is 0.08 °C/mm. From the above mentioned results, it is observed that addition of insulating materials like flyash and perlite in the SCC mix, does not cause a temperature gradient equal to that reported in Arlington road test.

For the Case 2, corresponding to minimum temperature, Arlington road test reported a temperature differential of −4.9 °C while the corresponding temperature gradient developed is −0.03 °C/mm. The negative sign indicates the curl up condition of the slab, indicating the lower part of the slab is at higher temperature than the surface. In the present study, the temperature gradient for Mix C is approximately the same as the reported value. This is attributed to the insulating property of perlite and flyash that have been added to the mix. Mix A and B show relatively lower magnitude of temperature gradient as developed in the Arlington road test.

The night time temperature gradients are considered as half the day time temperature gradients (Huang 2004). In the present study, the night time temperature gradients are around 30–40% the day time values for all the three mixes in Case 1. However, the night time gradients are in the range of 45–65% the day time gradient values for all three mixes in Case 2.

In the present study, a third case representing 24-h precipitation has been reported. It is observed that due to continuous precipitation, the overall temperature values are low. No significant temperature difference and gradient is developed. This is attributed to the moisture flow within the concrete. Water is 25 times more conductive than air (Demirboğa and Gul 2003). The presence of moisture in the pores of the concrete increases its thermal conductivity. Hence the temperatures are more or less constant throughout the 15 cm depth of the cube during 24-h precipitation. The temperature gradient developed in all the three mixes is insignificant for day time as well as night time.

Table 6. ΔT (°C) and ΔT/d (°C/mm) values for Mix A, B and C corresponding to the three cases

Sr. No.	Mix details	Case 1				Case 2				Case 3			
		Day		Night		Day		Night		Day		Night	
		ΔT (°C)	ΔT/d (°C/m m)	ΔT (°C)	ΔT/d (°C/m m)	ΔT (°C)	ΔT/d (°C/m m)	ΔT (°C)	ΔT/d (°C/m m)	ΔT (°C)	ΔT/d (°C/m m)	ΔT (°C)	ΔT/d (°C/m m)
1	A	5.56	0.045	−2.09	−0.017	4.9	0.039	−3.1	−0.025	1.02	0.008	−1.08	0.0086
2	B	7.25	0.058	−2.29	−0.019	7.75	0.062	−3.49	−0.028	0.06	0.0004 8	−0.4	0.0032
3	C	7.45	0.06	−2.72	−0.02	7.24	0.058	−3.63	−0.029	0.04	0.0003 2	−0.48	0.0038

6 Conclusions

1. Powder based mix design of SCC delivers a cohesive, robust concrete mix with minimum segregation and bleeding.
2. The temperature gradient developed in Mix A for Case 1 is much less (0.045 °C/mm) as compared to the result reported in Arlington Road Test (0.08 °C/mm).
3. For the minimum temperature condition (Case 2), Mix C develops a temperature gradient (−0.029 °C/mm) equivalent to the gradient observed in Arlington Road Test (−0.03 °C/mm). This is attributed to the insulating properties of flyash and perlite in the mix.
4. 24-h precipitation does not develop any significant temperature gradient.
5. Of all the three mixes, Mix A exhibits better thermal performance. This is evident from lower thermal gradient values developed for Mix A as compared to the other two mixes.
6. Due to high cement content and use of HRWR superplasticizer, SCC is expensive as compared to the normal concrete. Flyash addition as cement replacement can help reduce the production cost. However, due to insulating nature of flyash, the temperature gradient developed for flyash based mix (Mix B) is higher than the values observed for Mix A. As compared to Mix C, the thermal gradient values for Mix B are relatively low. Hence, from economy point of view, Mix B is recommended.

References

Chai, G., van Staden, R.: Impact of climate related changes in temperature on concrete pavement: a finite element study. In: 25th ARRB Conference Perth, Australia, Sept 23–26, 2012, pp. 1–12 (2012). ISSN 0572-1431

Delatte, N.: Concrete Pavement Design, Construction and Performance. (2008). ISBN 10: 0-203-96108-0

Demirboğa, R., Gul, R.: The effects of expanded perlite aggregate, silica fume and flyash on the thermal conductivity of lightweight concrete. Cem. Concr. Res. (2003). https://doi.org/10.1016/S0008-8846(02)01032-3

Domone, P.: Self-compacting concrete—an analysis of 11 years of case studies. Cem. Concr. Compos. (2006). https://doi.org/10.1016/j.cemconcomp.2005.10.003

EFNARC: The European Guidelines for Self-Compacting Concrete—Specification, Production and Use. European Federation of National Associations representing producers and applicators of specialist building products for concrete (2005)

Fwa T.F.: The Handbook of Highway Engineering (2006). ISBN 0-8493-1986-2

Gaimster, R., Dixon, N.: Self-compacting concrete. In: Newman, J., Choo, B.S. (eds.) Advanced Concrete Technology-Processes (2003). ISBN 0 7506 5105 9

Gandage, A., et al.: Optimization of class C flyash dosage in self compacting concrete for pavement application. In: Proceedings of International Conference on Innovations in Concrete 2013, Hyderabad, Andhra Pradesh, India, Oct 23–26, 2013 (2013). ISBN 978 81 7371 912

Gandage, A., et. al.: Effect of Perlite on Thermal Conductivity of Self Compacting Concrete (2013). https://doi.org/10.1016/j.sbspro.2013.11.111

Goodier, C.: Development of self-compacting concrete. Proc. Inst. Civ. Eng. **156**, 405–414 (2003)

Huang, Y.: Pavement Analysis and Design (2004). ISBN 0-13-142473-4

Kodur, V., Sultan, M.: Effect of Temperature on Thermal Properties of High Strength Concrete (2003). ISSN 0899-1561/2003/2-101–107

Kim, K.-H., Jeon, S.-E., Kim, J.-K., Yang, S.: An Experimental Study on Thermal Conductivity of Concrete (2003). PII: S0008-8846(02)00965-1

Rao, S., Roesler, J.: Characterizing effective built-in curling from concrete pavement field measurements. J. Transp. Eng. **131**(4), 320 (2005). https://doi.org/10.1061/(asce)0733-947x (2005)

Türkel, S., Kandemir, A. Fresh and hardened properties of SCC made with different aggregate and mineral admixtures (2010). https://doi.org/10.1061/(asce)mt.1943-5533.0000107CE

Static Compaction Characteristics of Coarse and Fine Grained Soils

Binu Sharma[1(✉)], Biplab Gogoi[2], and Asuri Sridharan[3]

[1] Department of Civil Engineering, Assam Engineering College, Guwahati, India
binusharma78@gmail.com

[2] Assam Engineering College, Guwahati, India
biplabgogoi20@gmail.com

[3] Indian National Science Academy, New Delhi, India
sridharanasuri@yahoo.com

Abstract. The modified Proctor's test, the reduced modified Proctor's test, the standard Proctor's test and the reduced standard Proctor's test are dynamic methods which are often used in the laboratory to ascertain the compaction characteristics of soils at different energy levels. To perform these tests considerable time and effort are needed. A laboratory procedure is devised to determine the relationships between moisture content and dry unit weight by using static compaction method for different static pressures. The static compaction pressure test is performed in the Proctor mould to statically compact the soil at different moisture contents. This method is less laborious and the time involved is less. An attempt was made to predict the compaction characteristics by static compaction method and parabolic curves were obtained between moisture content versus dry unit weight at different static pressures. The static pressure equivalent to the standard energy input of a standard Proctor test to obtain the Proctor's optimum moisture content (OMC) and maximum dry unit weight (MDUW) was determined for fine grained soil. The main objective of present study is to determine the static compaction characteristics of both coarse and fine grained soils and compare it with that of the dynamic compaction characteristics at different compactive efforts. For this, eleven different soil samples of classification CI, CH, SC, SP and SM were tested. The analysis shows that the relation between moisture content and dry unit weight in static compaction for different static pressure is parabolic in nature for CI & CH class of soil. For SP class of soil, both the static and dynamic compaction curves show an undulatory pattern with maximum dry unit weight near dry and towards saturated condition. The dynamic compaction curve for both SC and SM class of soil is parabolic in nature. For SM class of soil, static compaction curve show a wavy pattern with maximum dry unit weight at dry and near saturated condition whereas for SC class of soil only one sided compaction part of the curve for the rising portion of the dry of optimum side was generated. In case of coarse grained soils, an equivalent static pressure, at which maximum dry unit weight at optimum moisture content can be obtained corresponding to different dynamic efforts, could not be determined as that of fine grained soils. Further, the static pressure—dry unit weight relationship for a particular moisture content was identified as a rectangular hyperbola.

Z. Hossain et al. (eds.), *Solving Pavement and Construction Materials Problems with Innovative and Cutting-edge Technologies*, Sustainable Civil Infrastructures,
https://doi.org/10.1007/978-3-319-95792-0_4

1 Introduction

Soil compaction is a process of mechanical improvement whereby soil particles are forced to pack more closely together by reducing air voids with the application of some mechanical forces like static or dynamic loads to attain a compacted soil mass. In Geotechnical testing laboratories, the standard ASTM D698-91(2012) and modified ASTM D1557-91(2012) Proctor compaction tests are most commonly used to determine the compaction characteristics for proper control over the field compaction. These dynamic compaction test are laborious and time consuming and also has some limitations in determining the value of maximum dry unit weight (MDUW) and optimum moisture content (OMC). One of these limitation is that dynamic compaction applies equivalent energy to all types of soil and the data is not distinctive to any particular soil. Also, in the field static roller machine is mainly used for compaction. Therefore, there is a gap between the laboratory test and field method to measure the value of MDUW. Again, Hafez et al. (2010) established that in compaction by Proctor method, the bottom layer obtains more energy as compared to the middle and upper layers. Thus the dynamic compaction will give each soil layer different amount of energy and it also affects the homogeneity of soil specimens. Therefore different compaction methods have been used in an attempt to come as close as possible to real field situations, and to reduce the specimen compaction time and effort. This work on static compaction is an effort in this direction.

2 Literature Review

The study of the compaction properties of soils through static and dynamic compaction methods were made by different research workers at different time. Literature pertaining to static compaction is very limited. Hogentogler (1937) was probably the first to mention very briefly about the static pressure equivalent to Proctor's compacting effort. He stated that the soil should be compacted at a static pressure of 130 psi (about 896 kN/m^2) to get the equivalent Proctor's compaction. Berenhard and Krynine (1952) compared the static and dynamic compaction efficiencies in their study. Reddy and Jagadish (1993) discussed about the relationship of compaction energy, dry density and optimum moisture content (OMC) by static compaction of soil. Mesbah et al. (1999) introduced a quasi-static compaction technique. Hafez et al. (2010) introduced a new laboratory compaction method known as static packing pressure test which is designed as a static compaction technique. Furthermore Kenai et al. (2006), Yuce and Kayabali (2010), Dario (2011) etc. are some researchers in this direction.

Sharma et al. (2016) have devised a laboratory static compaction technique and attempted to predict the compaction characteristics by static compaction method; parabolic curves were obtained between moisture content versus dry unit weight at different static pressures. They also arrived at an equivalent static pressure corresponding to standard Proctor compaction energy in which standard Proctor maximum dry unit weight value can be obtained. They found this equivalent static pressure to be around 820 kN/m^2 corresponding to standard Proctor compaction energy. This finding was for fine grained soils having classification of CH, CL and ML according to the

Unified Soil Classification System (USCS). Sharma and Deka (2016) further extended the work on static compaction test of fine grained soils with four different dynamic compaction energies. However, the work was confined only to fine grained soils. The main objective of this study is to determine the static compaction characteristics of both coarse and fine grained soils and compare it with that of the dynamic compaction characteristics at different compactive energies.

3 Test Program

Altogether eleven soil samples were tested having soil classification of CH, CI, SC, SM and SP as per Indian Standard Soil Classification System (ISSCS). Table 1 shows the relevant physical properties of the soil samples. In the present study coarse grained soils of SC and SM classification are artificially prepared by mixing sand and fine grained soil in different proportion. The modified Proctor (MP), the reduced modified Proctor (RMP), the standard Proctor (SP) and the reduced standard Proctor (RSP) tests were used to determine dynamic compaction characteristics of soil. The SP test is done as per IS: 2720-part vii (1980) and the MP test is conducted as per IS: 2720-part viii (1983). The procedure and equipment for RSP and RMP tests are essentially same as that used for SP and MP test respectively; however each layer received 15 numbers of blows of a rammer instead of 25.

Table 1. Physical properties of soil samples

Sample No.	% of sand	% of (Silt + Clay)	Specific gravity (G)	Liquid limit (%)	Plastic limit (%)	IS classification
1	19.73	80.27	2.72	48.81	18.28	CI
2	21.60	78.40	2.65	57.40	16.36	CH
3	66.8	33.2	2.69	23.45	15.38	SC
4	62	38	2.70	36.62	18.52	SC
5	57.4	42.6	2.70	39.34	20.24	SC
6	77.6	22.4	2.63	–[a]	–[a]	SM
7	84.8	15.2	2.65	–[a]	–[a]	SM
8	90.4	9.6	2.66	–[a]	–[a]	SM
9	98	2	2.63	–[a]	–[a]	SP
10	96.4	3.6	2.64	–[a]	–[a]	SP
11	99	1	2.64	–[a]	–[a]	SP

[a]Non-plastic soil

The laboratory method of carrying out the static compaction test consists of placing a known weight of soil with known moisture content into the standard Proctor mould of 1000 cc capacity and then compacting it statically in a loading frame. The effect of dry unit weight with the variation of height of soil ranging from 25 to 100 mm in the Proctor mould was studied by Sharma et al. (2016) and it was found that there was no significant variation in dry unit weight corresponding to different static pressures.

Therefore in the present study, the static compaction test was performed in only one soil thickness of 100 mm. Two metal plates of diameter 98 mm and thicknesses 5 and 16 mm respectively were placed one above the other, on top of the soil sample in the mould. The entire assembly was placed under a cylindrical plunger of diameter 50 mm of the loading frame. Load was then statically applied to the soil through a proving ring having a proving ring constant of 0.99 kg/division. The height of penetration of the metal plate from the top surface of the mould was measured corresponding to different load levels. Since the soil was filled at known moisture content, the corresponding dry unit weight of the soil gets determined. The static pressure was gradually increased till no further increase in dry unit weight occurred with further increase in static pressure.

4 Experimental Results

In the static compaction test, for a particular moisture content the soil was subjected to different static pressures and the dry unit weights were calculated. The relationship between static pressure and dry unit weight for all the tested soil samples corresponding to different moisture contents have been plotted in the form of curves. Typical plot of static pressure versus dry unit weight for sample 3 at different moisture contents is shown in Fig. 1. It was observed that at lower static pressure, a significant variation in dry unit weight was obtained, but this variation became negligible with increase in static pressure and after a particular static pressure the dry unit weight became constant. This particular static pressure at which maximum dry unit was obtained, changed with the change in class of coarse grained soils. For sample SC, SM and SP classes of soils, this particular static pressures was obtained as 6500, 3000 and 6000 kN/m^2 respectively. In case of fine grained soils, this particular static pressure after which dry unit weight became constant, remained constant around 1513 kN/m^2, irrespective of the types of the fine grained soils. Similar finding on fine grained soils was also noticed by Sharma et al. (2016) in their study.

5 Analysis of Compaction Characteristics

The relationship between moisture content and dry unit weight for all the soil samples at different static pressures are determined and are represented in the form of curves. Superimposing both the static and dynamic compaction curves for moisture content versus dry unit weight relation, the results of typical curves for each class of soil are shown in Fig. 2 and from Figs. 6, 7 and 8. For fine grained soils both the static and dynamic compaction curves are parabolic in nature; but contrast to the conventional parabolic relationship of moisture content versus dry unit weight, static compaction curves of coarse grained soils showed a different pattern which also got changed with change in class of coarse grained soils. These are separately discussed below for the different classes of soils. Determination of equivalent static pressures corresponding to different dynamic compaction effort are also discussed for the different soil classification considered in the study.

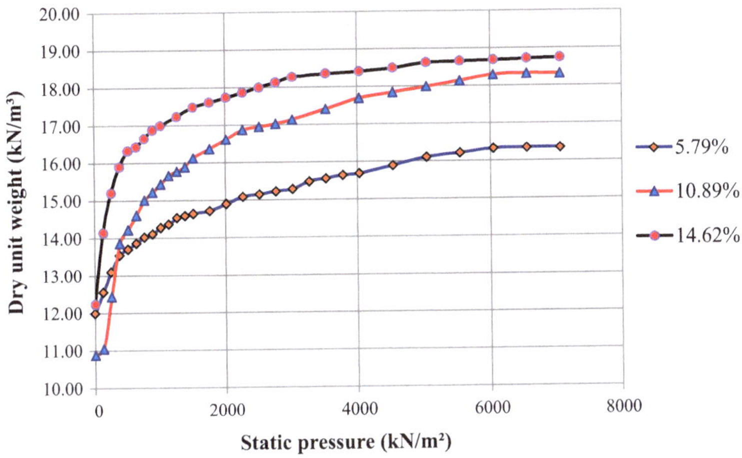

Fig. 1. Static pressure versus dry unit weight curve of sample 3 at different moisture contents

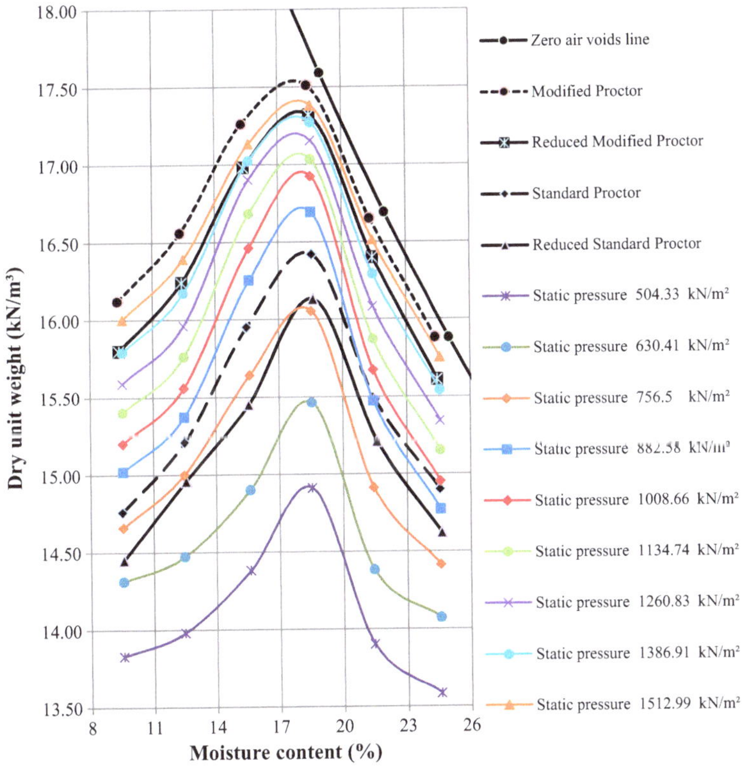

Fig. 2. Static and dynamic compaction curves of sample 1 (Soil classification CI)

5.1 Fine Grained Soils (Sample 1 and 2)

For determining the compaction characteristics of fine grained soils both static and dynamic compaction methods are used. Dynamic compaction test results for the MP, RMP, SP and the RSP tests results are represented in the form of curves, which are parabolic in nature. The maximum dry unit weight (MDUW) obtained for the two tested fine grained soil samples corresponding to the MP, RMP, SP and the RSP tests lies between 16.95 and 18.24 kN/m^2, 16.75 and 18.04 kN/m^2, 16.12 and 17.23 kN/m^2 and 15.87 and 16.92 kN/m^2 respectively. From the static compaction test results when curves are plotted between the moisture content versus dry unit weight at different static pressures, similar parabolic curves are obtained as that of dynamic compaction. A typical result of the superimposed static and dynamic compaction curves for sample 1 is shown in Fig. 2. For the two tested fine grained soil samples, it was noticed that the MDUW obtained from static compaction test corresponding to the maximum static pressure of 1513 kN/m^2 lies in between the MDUW value obtained from the MP and the RMP compaction tests. So equivalent static pressures can be derived to obtain the MDUW value corresponding to RMP, SP and RSP compaction tests. Again, for the two tested fine grained soil samples the MP compaction curves always lie above the static compaction curves corresponding to the maximum static pressure; so it is not possible to obtain the equivalent static pressure corresponding to the MP test as the MDUW value obtained from MP test is more than that of the MDUW value obtained from static compaction test at the maximum static pressure of 1513 kN/m^2.

5.1.1 Determination of Equivalent Static Pressures for Fine Grained Soils

Equivalent static pressure corresponding to a dynamic compactive energy is defined as that static pressure which should be applied to a statically compacting soil sample to obtain the same maximum dry unit weight (MDUW) value at OMC as that is obtained by the dynamic compactive energy. For determining the equivalent static pressure corresponding to the SP test, two maximum dry unit weight (MDUW) corresponding to two static pressures are considered in such a way that one MDUW is above the MDUW obtained from SP test and the other is below it. Assuming linear variation of dry unit weights with the two pressures, the pressure corresponding to the SP tests maximum dry unit weight was determined. Similar method was applied to find the equivalent static pressure for RSP test and RMP test respectively. A typical graphical representation for the determination of equivalent static pressure corresponding to RMP test is shown in Fig. 3.

The results of equivalent static pressures are shown in Table 2. The average equivalent static pressures come out to be around 836, 785 and 1427.5 kN/m^2 corresponding to the SP, RSP and RMP tests respectively. Hence it can be concluded that in static compaction in case of fine grained soils, when a static pressure of around 836 kN/m^2 is applied, a static compaction curve, equivalent to the SP curve can be obtained. Similar can be said for the RSP test and RMP test.

In the study on the compaction characteristic of fine grained soil by Sharma et al. (2016) an equivalent static pressure of around 820 kN/m^2 for the SP test was determined; and again in the study by Sharma and Deka (2016) the average equivalent static

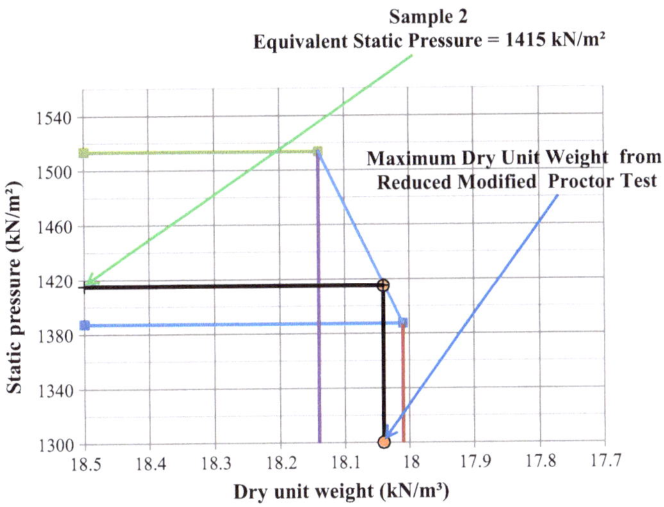

Fig. 3. Determination of equivalent static pressure of sample 2 for reduced modified Proctor test

Table 2. Equivalent static pressure (ESP) for the fine grained soil samples

Sample No.	ESP for standard Proctor test (kN/m²)	ESP for reduced standard Proctor test (kN/m²)	ESP for reduced modified Proctortest (kN/m²)
1	830	770	1440
2	842	800	1415
Average ESP	836	785	1427.5

pressures came out to be around 845, 788 and 1426 kN/m^2 for SP test, RSP test and RMP test respectively, which is in agreement with the present work.

Taking the data from this work and from Sharma and Deka (2016), an attempt is also made to see the relationship between the MDUW and the OMC obtained from the static compaction test at equivalent static pressures and that obtained from the dynamic compaction test for the three different compactive energies. The relationship between the MDUW corresponding to the equivalent static pressure of 845, 788 and 1426 kN/m^2 (Sharma and Deka 2016), and around 836, 785 and 1427.5 kN/m^2 (present work) for SP test, RSP test and RMP test respectively and the maximum dry unit weight from the three Proctor tests is shown in Fig. 4. This relationship is given by Eq. 1.

$$(\text{MDUW})_{\text{static}} = 1.01\,(\text{MDUW})_{\text{Proctor}} - 0.177 \qquad (1)$$

with a high correlation coefficient of 0.948.

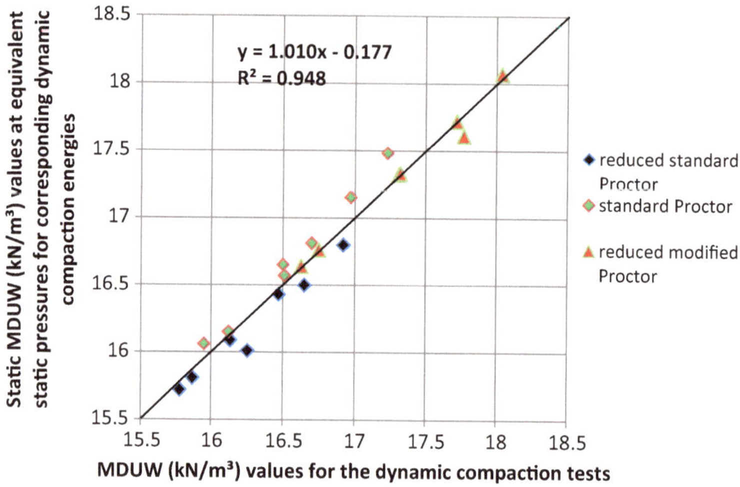

Fig. 4. Plots of MDUW (kN/m^3) obtained by the static pressure curves corresponding to the equivalent static pressures and MDUW (kN/m^3) obtained by the three Proctors test

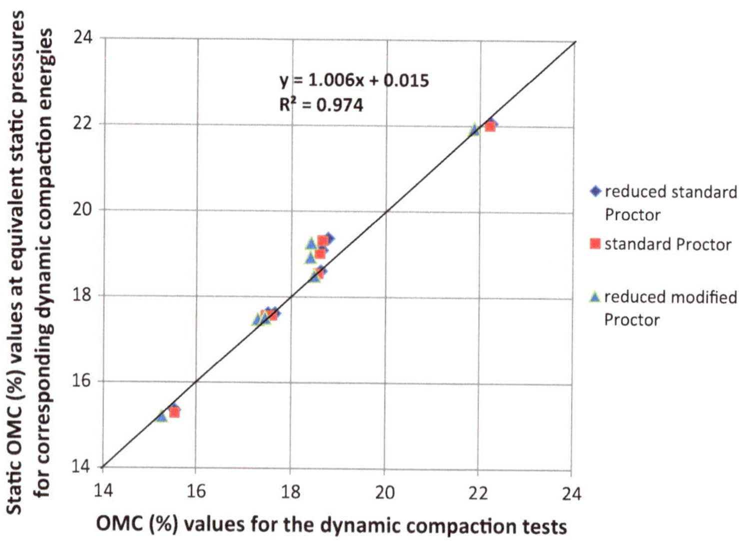

Fig. 5. Plots of OMC obtained by the static pressure curves corresponding to the equivalent static pressures and OMC obtained by the three Proctors test

Similarly, the relationship between the OMC corresponding to the equivalent static pressures and the OMC from the Proctor tests is shown in Fig. 5. This relationship is given by Eq. 2 as

$$(\mathrm{OMC})_{\mathrm{static}} = 1.006(\mathrm{OMC})_{\mathrm{Proctor}} + 0.015 \quad (2)$$

with a correlation coefficient of 0.974.

The above Eqs. 1 and 2 show the close correlation between the dynamic compaction values and the static compaction values with static compaction done with the respective equivalent static pressures.

5.2 Coarse Grained Soils of Classification SC (Samples 3, 4 and 5)

For coarse grained soils having soil classification SC, the dynamic compaction curves for MP, RMP, SP and RSP tests are parabolic in nature. For the three tested SC class of soils, the maximum dry unit weight (MDUW) value obtained from dynamic compaction test at different compactive efforts lie between 18.05 and 19.61 kN/m^3. In static compaction test, only the one sided compaction curves at different static pressure for the rising portion of the dry of optimum side were generated as shown in Fig. 5. It is not possible to obtain the drooping portion beyond the OMC normally noticed in the conventional Proctor curves. When the moisture content was increased beyond the OMC to a soil sample and then statically compacted, water drained out during compaction and at a particular static pressure penetration ceased. At this point moisture content was found to again correspond to the optimum moisture content (OMC) and the same dry unit weight was obtained which was already obtained corresponding to the OMC value. Again, it is not necessary to report the compaction beyond OMC since water drained out during the process and it is not a purely compaction phenomenon. The MDUW value obtained corresponding to the maximum static pressure of 6500 kN/m^2 is less than the MDUW value obtained from the SP test due to which no equivalent static pressure could be determined. These one sided static compaction curves were also noticed by Reddy and Jagadish (1993) in their study of static compaction test in a soil sample having percentage of fine 51.2%.

It is observed in the case of fine grained soils, parabolic compaction curve is obtained under static compaction at different static pressures. As the soil changes to SC category, only one sided compaction curve for dry of optimum is obtained under static compaction. The transition phase of the parabolic compaction curve of fine grained soil to the one sided compaction curve of SC category soil in static compaction need to be further investigated.

5.3 Coarse Grained Soils of Classification SM (Sample 6, 7 and 8)

The dynamic compaction curves obtained from the MP, RMP, SP and RSP tests for the SM class of soils are also parabolic in nature. For the tested soil samples, the maximum dry unit weight (MDUW) value obtained from dynamic compaction test at the different compactive energies lie between 18.17 and 20.39 kN/m^3. Static compaction curves show a wave like pattern, where additional to MDUW at OMC, MDUW also can be

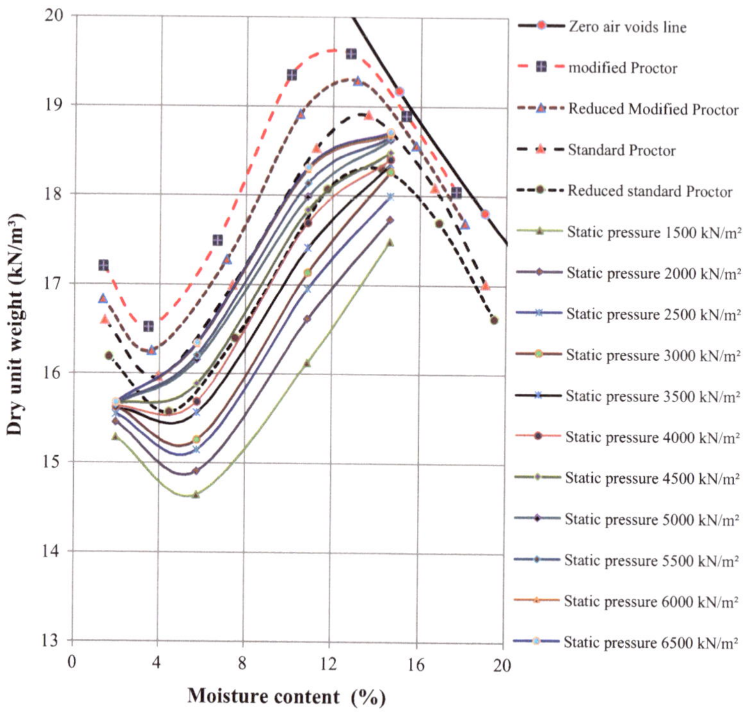

Fig. 6. Static and dynamic compaction curves of sample 3 (Soil classification SC)

achieved at dry and near saturated condition of soils as shown in Fig. 6. For the three SM class of tested soil samples, the MDUW values obtain in the static compaction test corresponding to the maximum static pressure of 3000 kN/m^2 lie between 16.77 and 17.02 kN/m^3 which are less than the MDUW value obtained from RSP compaction test. So no equivalent static pressure can be obtained. This is mainly due to the variation in the arrangement of soil structure under static and dynamic loading. In dynamic compaction, soil particles have more degree of freedom to rearrange amongst themselves to attain a densely pack structure as compared to static compaction where soil particles have less freedom to rearrange amongst themselves.

5.4 Coarse Grained Soils of Classification SP (Sample 9, 10 and 11)

For SP class of soils, both the dynamic compaction curves at the four different compaction energies and static compaction curves at different static pressures exhibit an undulatory pattern as shown in Fig. 7. Regardless of whether it is static compaction or dynamic compaction, MDUW is always noticed towards dry condition. Between a moisture content of 2–5% there is minimum dry unit weight state, which is due to the bulking phenomena of sand. MDUW also can be obtained towards saturated side. In case of sand it is always difficult to determine the OMC value as the compaction curve contains more than one peak. It is always recommended to compact sand either in dry state or in saturated state by flooding with water. For all the three SP class of soil samples, the maximum dry unit weight (MDUW) value obtained from dynamic compaction test at different compactive energies lie between 16.20 and 17.92 kN/m^3; whereas the MDUW values obtain in the static compaction test corresponding to the maximum static pressure of 6000 kN/m^2 lie between 14.75 and 14.86 kN/m^3. It is noticed that static effort is very less effective as compared to dynamic effort in com-

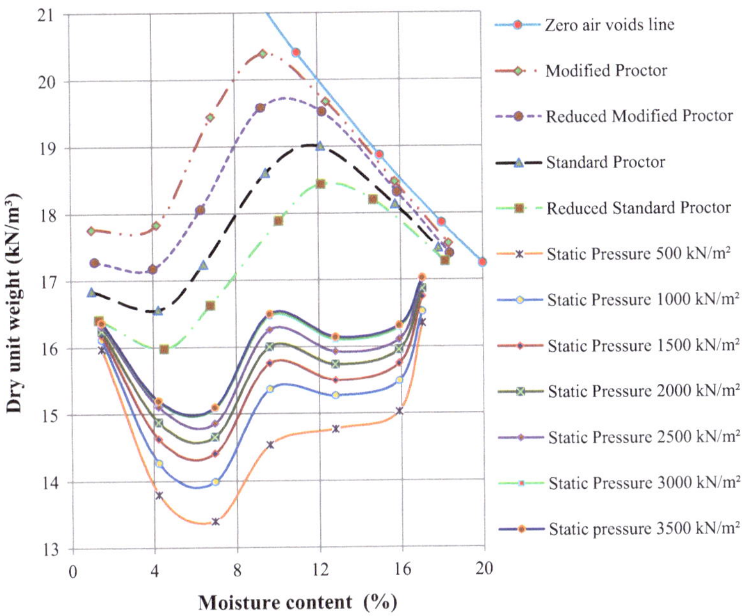

Fig. 7. Static and dynamic compaction curves of sample 6 (Soil classification SM)

paction of pure sand, as MDUW obtained by static compaction corresponding to the maximum static pressure is comparatively very less than that of dynamic compaction; this is mainly due to the variation in soil structure attained by both the methods. However in attaining a maximum density state both static and dynamic methods are ineffective; in real situation alternate compaction techniques like vibratory compaction, tamping etc. give much higher density state in case of sand (Fig. 8).

It is noticed from the test results that in the case of coarse grained soils the maximum dry unit weight (MDUW) value obtained by the static compaction method at maximum static pressure is less than the MDUW value obtained by the standard Proctor method. However the difference in the MDUW value obtained by both the methods goes on decreasing with the increase in percentage of fines in coarse grained soils and in the case of fine grained soils, the static MDUW value corresponding to maximum static pressure is greater than that of dynamic MDUW value corresponding to standard Proctor compaction. In the case of static loading of coarse grained soils, readjustment of structure by particle orientation is less. Whereas in fine grained soils, small matrices are formed leaving large voids between the matrices that can be closed by static loading. So in the case of SM and SP type of soils, it is better to compact it with vibratory compaction or by tamping. The static compaction characteristics of

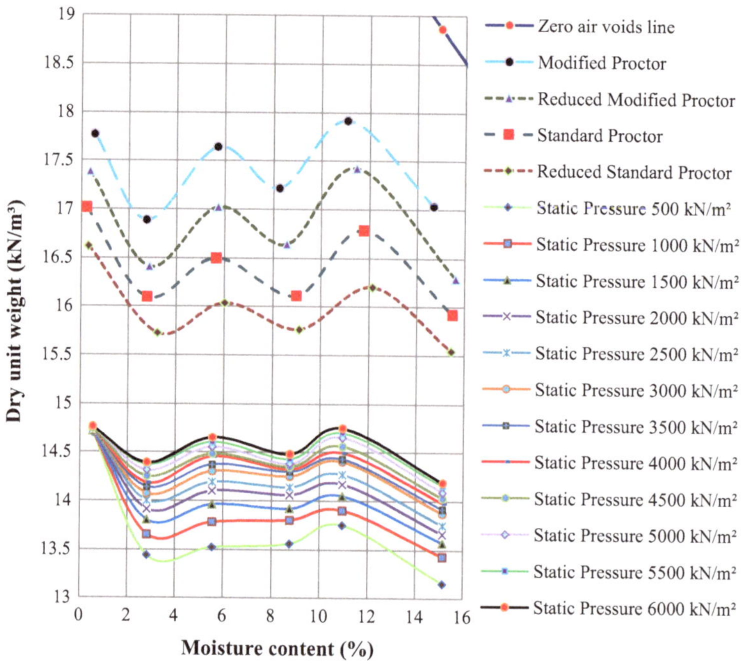

Fig. 8. Static and dynamic compaction curves of sample 9 (Soil classification SP)

coarse grained soil is totally different from that of fine grained soils; no similar parabolic curve can be obtained for moisture content versus dry unit weight relationship at different static pressures and no generalized equivalent static pressure can be determined as that was obtained for fine grained soils.

6 Conclusions

In the relationship between static pressure and dry unit weight obtained from static compaction test, it was observed that at lower static pressure, a significant variation in dry unit weight was obtained, but this variation became negligible with increase in static pressure and after a particular static pressure the dry unit weight became constant. This particular static pressure got changed with the change in class of coarse grained soils. But, in the case of fine grained soils, it was found that this particular static pressure remained constant irrespective of the types of the fine grained soils. For fine grained soils both the static and dynamic compaction curves are parabolic in nature. Compaction curves of coarse grained soils showed a different pattern which also changed with change in class of coarse grained soils. For SC class of soils, the dynamic compaction curves are parabolic in nature whereas in static compaction, only the one sided compaction curves for the rising portion on the dry of optimum side was generated. For SM class of soils, the dynamic compaction curves are parabolic in nature and static compaction curves show a wave like pattern. For SP class of soils, both the static and dynamic compaction curves exhibit an undulatory pattern. In the case of coarse grained soils the MDUW value corresponding to the maximum static pressure obtained by the static compaction method is less than the MDUW value obtained by the standard Proctor method and in the case of fine grained soil the static MDUW value corresponding to the maximum static pressure is greater than that of MDUW corresponding to standard proctor compaction. For fine grained soils, equivalent static pressures corresponding to standard Proctor, reduced Standard Proctor and Reduced Modified Proctor compaction efforts are around 836, 785 and 1427.5 kN/m^2 respectively; and equivalent static pressure cannot be achieved corresponding to modified Proctor's compaction effort. For coarse grained soils, no generalized equivalent static pressures can be determined.

References

ASTM D698–91: Test Method for Laboratory Compaction Characteristics of Soil using Standard Effort [12,400 ft-Ibf/ft3 (600 kN-m/m3)]. ASTM International, West Conshohocken, PA. www.astm.org (2012)

ASTM D1557–91: Test Method for Laboratory Compaction Characteristics of Soil using Modified Effort [50,400 ft-lbf/ft3 (2,700 kN-m/m3)]. ASTM International, West Conshohocken, PA. www.astm.org (2012)

Berenhard, R.K., Krynine, D.P.: Static and dynamic soil compaction. In: Proceedings of the 31st Annual Meeting of the Highway Research Board, Washington, D.C., January, TRB, Washington, D.C., Unpublished (1952)

Bureau of Indian Standards: Methods of test for soils: determination of water content-dry density relation using Light Compaction. IS 2720-7, New Delhi, India (1980)

Bureau of Indian Standards: Methods of test for soils: determination of water content-dry density relation using Heavy Compaction, IS 2720-8. New Delhi, India (1983)

Dario, C.: The influence of laboratory compaction methods on soil structure: mechanical and micro-morphological analysis. Soils Rocks, Sao Paulo **34**(1), 91–98 (2011)

Hafez, M.A., Doris Asmani, M., Nurbaya, S.: Comparison between static and dynamic compaction methods. Electron. J. Geotech. Eng. (EJGE) **15**, 1641–1650 (2010)

Hogentogler, C.A.: Engineering Properties of Soil. McGraw-Hill Book Company Inc, New York (1937)

Kenai, S., Bahar, R., Benazzoug, M.: Experimental analysis of the effect of some compaction methods on mechanical properties and durability of cement stabilized soil. J. Mater. Sci. **2006** (41), 6956–6964 (2006)

Meshab, A., Morel, J.C., Oliver, M.: Clayey soil behaviour under static compaction test. Mater. Struct. **32**(9), 687–694 (1999)

Reddy, B.V.V., Jagadish, K.S.: The static compaction of soils. Geotech. J. **43**(2), 337–341 (1993). The Institution of Civil Engineers London

Sharma, B., Deka, A.: Static compaction test and determination of equivalent static pressure. In: Proceedings of I.G.C., 2016, December 15–17, 2016, IIT Madras, Chennai, India, pp. 1–4 (2016)

Sharma, B., Sridharan, A., Talukdar, P.: Static method to determine compaction characteristics of fine-grained soils. Geotech. Test. J. (2016). ASTM. https://doi.org/10.1520/gtj20150221

Yuce, E., Kayabali, K.: The use of static compaction method. Geological Congress of Turkey, Book of Abstracts, pp. 174–175 (2010)

Resilient Modulus and Layer Coefficient of Open-Graded Aggregates

Hung Tan Nguyen, Aryssa Kathreen B. Marcaida, and Jaehun Ahn(✉)

Department of Civil and Environmental Engineering, Pusan National University, Busan, South Korea
nthung010189@gmail.com, jahn@pusan.ac.kr

Abstract. In permeable pavement, base and subbase layers need to meet design requirements capable of supporting traffic loads and providing adequate drainage for road serviceability. With high porosity in structure, open-graded aggregates are considered as a general suitable material for base and subbase layers. Nevertheless, because of few fine-grained particles and high void ratio in open-graded aggregates, load-bearing capacity of pavement may be reduced. Resilient modulus and layer coefficient are two necessary parameters in structural design of pavement. In decades, there were few studies made for base and subbase layers made from open-graded aggregates. The overall objective of this study is to summarize different researches made about the resilient modulus and layer coefficient of permeable base and subbase layers.

Keywords: Open-graded aggregate · Porous pavement · Resilient modulus Layer coefficient · Triaxial cell

1 Introduction

In areas where low impact development strategies have increased, the use of permeable pavement systems have been widely adopted for mitigating the effects of urbanization. In permeable pavement systems, base and subbase layers are not only responsible for spreading traffic load from pavement surface to the underlying subgrade, but also are allocated for specific functions. In decades, open-graded aggregate has been a popular material for base/subbase layers for permeable pavement due to its satisfactory performance on load-bearing and drainage, and affordability. However, there are few researches made about the performances of open-graded aggregate in pavement design system. In pavement design system, resilient modulus and layer coefficient are two important parameters. In this study, some factors that have significant influences on the resilient modulus and layer coefficient of open-graded aggregate are summarized and presented in Table 1.

Z. Hossain et al. (eds.), *Solving Pavement and Construction Materials Problems with Innovative and Cutting-edge Technologies*, Sustainable Civil Infrastructures,
https://doi.org/10.1007/978-3-319-95792-0_5

Table 1. Summary resilient modulus and layer coefficient of base/subbase material

Property	Material	Literature	Remark
Resilient modulus	Limestone, Gravel	Heydinger et al. (1996)	Table 3
	Aggregate	Zhou et al. (1993)	Figs. 1, 2 and 3
		Ba et al. (2011)	Figs. 4 and 5
Layer coefficient	Almost types of aggregate	Bahia et al. (2000)	Table 4
	Asphalt	Bahia et al (2000)	Table 5

2 Resilient Modulus of Open-Graded Aggregate

Huang (2004) defined that resilient modulus, represented as M_r, is a parameter that is determined from the ratio of the repeated deviator stress to the recovery strain. In designing pavement, resilient modulus is used for characterizing the non-linear stress behavior of base aggregates and soil subgrades subjected to traffic loadings.

$$M_r = \frac{\sigma_d}{\varepsilon} \tag{1}$$

where:
M_r = resilient modulus (kPa)
σ_d = repeated deviator stress (kPa)
ε = recoverable strain.

There are many factors that influence the resilient modulus of open-graded aggregate material. Summary of some significant factors are presented below:

Gradation and Type

Zhou et al. (1993) have researched the resilient modulus of six different aggregate gradations. Among them, open-graded aggregate named A is a material acquired locally, while aggregate B was from New Jersey. The gradation of these materials are presented in Table 2. Figures 1 and 2 show the results of resilient modulus test performed on both aggregates.

Table 2. Sieve analysis two gradation of open-graded aggregate (after Zhou et al. 1993)

Aggregate sieve size	Open-graded A	Open-graded B
1-½ "	100	100
1"	97.5	97.5
¾"	67.5	86
½ "	56.5	70
¼"	37.5	54
#10	7.5	12.5
#40	4	3
#200	1	1.5

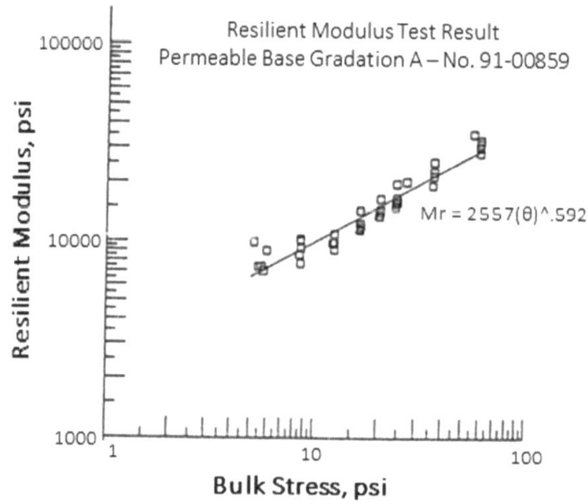

Fig. 1. Resilient modulus test for gradation A (Zhou et al. 1993)

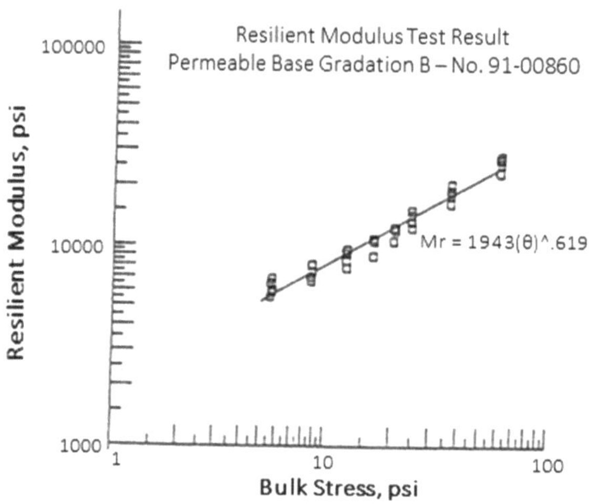

Fig. 2. Resilient modulus test for gradation B (Zhou et al. 1993)

The results show that there is a slight change in values of resilient modulus between two gradations. In case of open-graded aggregates, gradation A obtained a little higher value than gradation B. Reasonably, gradation has a small effect to resilient modulus.

Moreover, another factor that contributes to resilient modulus is the type of material. Different materials have different physical properties. In the research of Heydinger et al. (1996), they compared the resilient modulus of two types of open-graded aggregates such as limestone and gravel under both dry and moist conditions. The results of resilient modulus were shown in Table 3. According to the results, the moisture condition made a slight change to the resilient modulus. However, there is a

Table 3. Resilient modulus of limestone and gravel (after Heydinger et al. 1996)

Moisture condition	Limestone		Gravel	
	Range of M_r (MPa)	M_r* (MPa)	Range of M_r (MPa)	M_r* (MPa)
Dry	50.4–295.5	218.7	67.6–481.7	347.1
Moisture	51–307	216.5	68.3–453.2	354.2

M_r* is resilient modulus in for Sequence No. 11 with both confining and maximum deviator equal to 101.4 kPa (15 psi)

significant difference on the resilient modulus between two material types used. The test results for limestone is much lower than that for gravel.

Fractured faces

Aside from studying the effect of aggregate gradation to resilient modulus, Zhou et al. (1993) also conducted experiment on the influence of fractured faces of aggregates. The specimens used are the aggregates acquired from New Jersey and are classified based on two fractured faces percentage—88% for Aggregate B and 100% for Aggregate E. Clearly, the percentage of fractured faces has significant effect to the resilient modulus. In comparison, higher resilient modulus is found in higher fractured faces percentage as shown in Fig. 3.

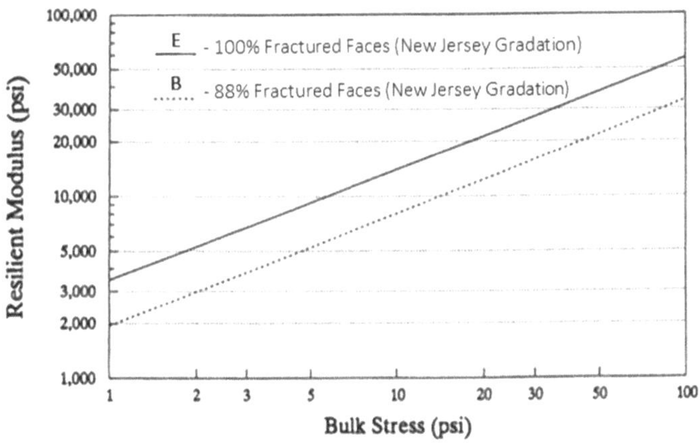

Fig. 3. Effect of fractured faces on resilient modulus for open-graded aggregate (Zhou et al. 1993)

State of stress

Another important factor during experiment is state of stress. Hicks and Monismith (1971) reported that with aggregate materials, resilient modulus increases with increasing confining stress. Ba et al. (2011) conducted experiments with unbound aggregate materials with different confining pressure. The aggregates were tested are native to Senegal which named GRB (red quartzite), GNB (Black quartzite), Basalte and Bandia limestone. By increasing the confining pressure with different types of aggregates and water content, the tests give remarkable results as shown in Figs. 4 and 5. Based on the figures, the resilient modulus has a nearly linear relationship with the confining pressure.

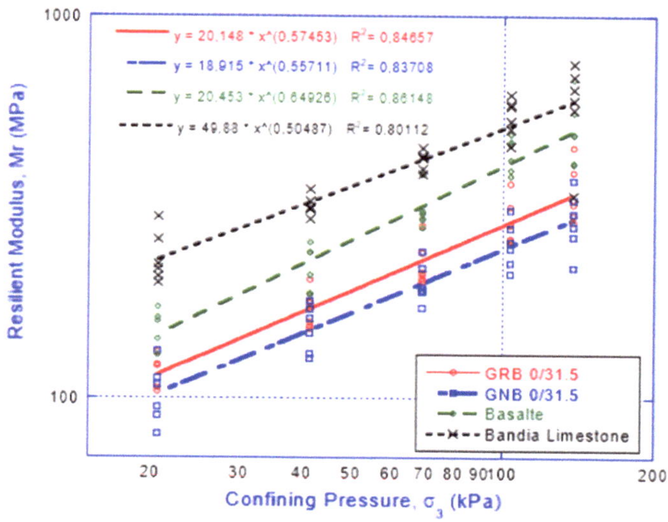

Fig. 4. Resilient modulus for different materials tested at optimum water content (Ba et al. 2011)

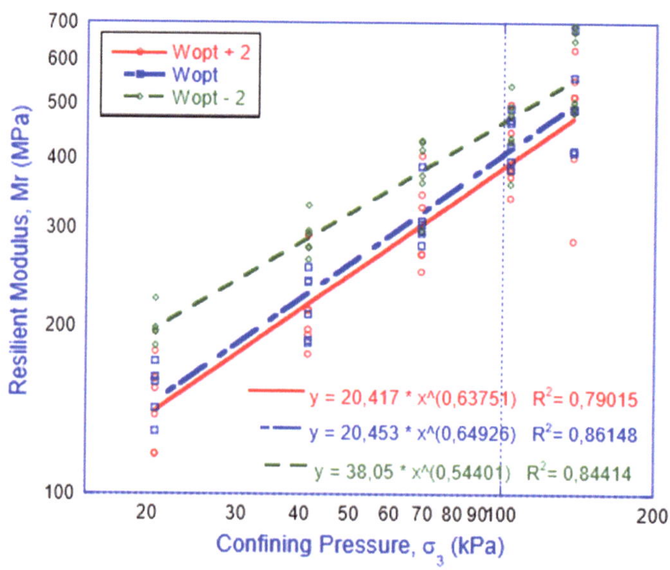

Fig. 5. Resilient modulus for basalt material with different water content (Ba et al. 2011)

Resilient modulus of open-graded aggregate is being researched a few in recent. It depends on many factors such as gradation, fractured faces, state of stress and type of material. Among these, the state of stress and type of material are the two factors that significantly affect resilient modulus.

Furthermore, the shape of particles also plays an important factor which was researched by a number of researchers. Hicks (1970), Hicks and Monismith (1971), Allen and Thompson (1974), Thom (1988), Barksdale and Itani (1989), Thom and Brown (1989) have reported that the value of resilient modulus of aggregates with angular to sub-angular shape is higher than the one with sub-round or round shape.

3 Layer Coefficient of Open-Graded Aggregate

Layer coefficient a_i is a factor multiplied to the layer of a given material required to carry an expected load (Huang 2004). Layer coefficient is developed by the AASHTO Guide for Design of Pavement Structures as a multiplier for pavement structural design. According to Koroma (2011), this value is not only used for converting the layer thickness into structural numbers, but also represents the relative impact of the layer coefficient, thickness, and the drainage coefficients on the structural number, as given by the equation

$$SN = a_1D_1m_1 + a_2D_2m_2 + a_3D_3m_3 + \cdots + a_iD_im_i \quad (2)$$

where:

SN = structural number

a_i = layer coefficient of base or subbase course

D_i = thickness for layer i

m_i = drainage coefficient for layer i.

Rada and Witczak (1981) indicated that the layer coefficient values were also determined as functions of California Bearing Ratio, compaction, saturation and thickness of asphalt. The values for aggregate and treated drainage layer were reported by Zhou et al. (1993), from 0.08 to 0.14. In case of asphalt treated permeable base, values between 0.14 and 0.19 were assigned.

A report by Mathis (1989) showed that the load-bearing capacity of New Jersey and Pennsylvania untreated permeable material is similar to dense-graded aggregate base. Based on that, Zhou et al. (1993) recommended that the layer coefficient of open-graded and dense-graded aggregate are similar and can be used in pavement design system. It indicated that when designing the permeable base/subbase layers with open-graded aggregates, the layer coefficients can be implied. As a result, Bahia et al. (2000) reported the value in Table 4 can be used as a reference document.

Furthermore, by using field tests with Falling Weight Deflectometer (FWD), resilient modulus and layer coefficient of asphalt were summarized by Bahia et al. (2000) as shown in Table 5.

Table 4. Summary the results of Layer Coefficients by 1972 AASHTO Survey (after Bahia et al. 2000)

Material	Layer coefficient
Limestone	0.14
Sandstone	0.13
Granite	0.12
Coarse grade	0.3–0.34
Gravel	0.09–0.34
Crush stone	0.14–0.29
Crush rock	0.06–0.16
Granular	0.07
Aggregate	0.06–0.14

Table 5. Summary of M_r and a_i estimated from FWD results (after Bahia et al. 2000)

Material	Resilient modulus (MPa)	Layer coefficient a_i	Test temp (°C)	Wisconsin layer coefficient	
				1972	1997
SHRP	1645	0.336	15.6	0.44	0.44
MV	385	0.083	32.3	0.44	0.44
ABC	714	0.13	6.1	0.3	0.34
Pul	814	0.31	15.6	0.1	0.1–0.25

Layer coefficient is an important parameter for determining the structure and thickness of pavement system. Many factors are related to the structural integrity of pavement. Layer coefficient for permeable aggregate material is similar with dense-graded material based on many researches.

4 Experimental Analysis of Resilient Modulus of Mixture of Soil and Open-Graded Aggregates

For the experiment, specimens were prepared by compacting mixture of subgrade soil and open-graded aggregates using modified Proctor hammer. Subgrade soil was prepared with optimum water content and aggregates have the particle size distribution shown in Fig. 6.

The prepared specimen was assembled on the triaxial cell between two porous stones (Fig. 7). The triaxial cell is assembled to the material testing system (MTS) unit. The triaxial cell is subjected to apply the desired confining pressure to the specimen while the MTS is equipped with actuator capable to induce cyclic axial stress. Load cell, pressure sensor and two linear variable differential transformation (LVDTs) were

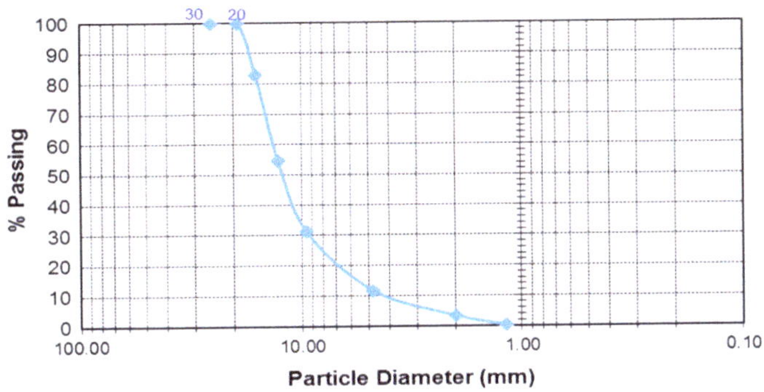

Fig. 6. Particle size distribution of aggregate

Fig. 7. Specimen in triaxial cell

connected to the computer for data recording. Data acquisition program was set up for collecting 280 readings per cycle (a reading records time, load values, confining pressure and deformation of two LVDTs simultaneously). Since there is no standard for testing mixture of soil and open-graded aggregate, in this case, the specimen was assumed to behave as subgrade soil. The resilient modulus test was conducted following the procedure NCHRP 1-28A for granular subgrade material in this study.

Figure 8 illustrates the results of resilient modulus test plotted against bulk stress. It shows that the values of resilient modulus increase with bulk stress increase. The test results agree with other researches that the relationship between the resilient modulus and bulk stress is proportional for different types of materials. In Fig. 9, the relationship between the octahedral stress and resilient modulus shows a similar linear behavior.

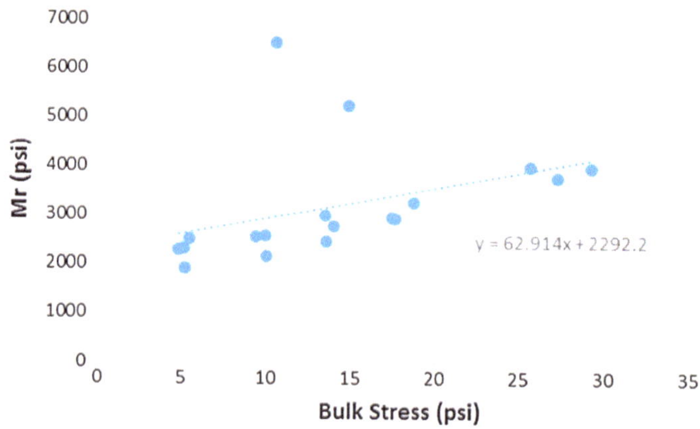

Fig. 8. Resilient modulus with bulk stress

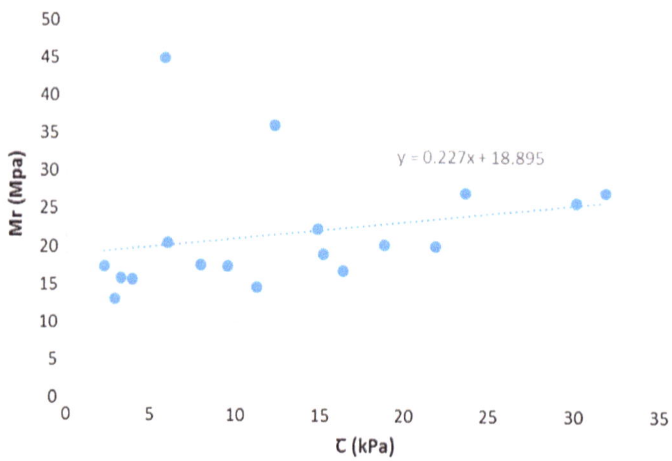

Fig. 9. Resilient modulus with octahedral stress

5 Conclusions

Resilient modulus and layer coefficient are two important parameters for open-graded aggregate which is used for base and subbase layers. Above discussion indicated that the layer coefficient of open-graded and dense-graded aggregates are similar. As a result, for open-graded aggregates, the layer coefficient can be evaluated from previous researches. On the other hand, resilient modulus of open-graded aggregates is a structural parameter which depends on many factors such as gradation, type of material, fractured face and state of stress among many other factors. The effect of material type on the resilient modulus of open-graded aggregates needs to be researched for better understanding of the characteristics of base and subbase layers in permeable pavement. Further, compaction of open-graded aggregates in laboratory is quite difficult because of its high porosity which reduces the effect of compaction. In future, procedure for compaction and resilient modulus test of aggregate with very high porosity need to be researched and developed for application on permeable pavement.

Acknowledgements. This research was supported by Korea Agency for Infrastructure Technology Advancement (KAIA) grant funded by the Ministry of Land, Infrastructure and Transport (Grant 18CTAP-C132363-02), and Basic Science Research Program through the National Research Foundation of Korea (NRF) funded by the Ministry of Education (No. NRF-2017R1D1A3B03034563).

References

Allen, J.J., Thompson, M.R.: Resilient response of granular materials subjected to time-dependent lateral stresses. Transp. Res. Rec. (510) (1974)

Ba, M., et al.: Resilient modulus of unbound aggregate base courses from Senegal (West Africa). Open J. Civil Eng. **1**(01), 1 (2011)

Bahia, H.U. et al.: Layer Coefficients for New and Reprocessed Asphaltic Mixes. University of Wisconsin, Madison, Department of Civil and Environmental Engineering, Madison, USA (2000)

Barksdale, R.D., Itani, S.Y.: Influence of aggregate shape on base behavior. Transp. Res. Rec. (1227) (1989)

Heydinger, A., et al.: Analysis of resilient modulus of dense-and open-graded aggregates. Transp. Res. Rec. J. Transp. Res. Board **1547**, 1–6 (1996)

Hicks, R.G., Monismith, C.L.: Factors influencing the resilient response of granular materials. Highway Res. Rec. (345) (1971)

Hicks, R.G.: Factors influencing the resilient properties of granular materials. Ph.D. thesis, University of California, Berkeley, Berkeley, USA (1970)

Huang, Y.H.: Pavement Analysis and Design. University of Kentucky, USA (2004)

Koroma, A.A.: Evaluation of the performance and costeffectiveness of pavement sections containing open-graded base courses. Master thesis, Michigan Technological University (2011)

Mathis, D.M.: Design and Construction of Permeable Base Pavement. Department of Transportation, FHWA, U.S (1989)

NCHRP, Laboratory Determination of Resilient Modulus for flexible Pavement Design.: National Cooperative Highway Research Program (NCHRP), Transportation Research Board of National Academies, Washington (2004)

Rada, G., Witczak, M.W.: Comprehensive evaluation of laboratory resilient moduli results for granular material (No. 810) (1981)

Thom, N.H., Brown, S.F.: The mechanical properties of unbound aggregates from various sources. In: Unbound Aggregates in Roads, pp. 130–142 (1989)

Thom, N.H.: Design of road foundations. Ph.D. thesis, Department of Civil Engineering, University of Nottingham, Nottingham, England (1988)

Zhou, H., et al.: Determination of free-draining base materials properties. Transp. Res. Rec. 1425, 54–63 (1993). Transportation Research Board, Washington, D.C.

Zhou, et al.: Determination of free-draining base materials properties. Transp. Res. Rec. 54–54 (1993)

Coefficient of Subgrade Reaction for the Permeable Block and Base System at Korea GI and LID Center

Yong-Jin Choi[1], Seungjun Lee[1], Jeongho Oh[2], Jaehun Ahn[1(✉)], and Jongwon Jung[3]

[1] Pusan National University, Busan, South Korea
{yj.choi,jahn}@pusan.ac.kr, juny9310@gmail.com
[2] Korea National University of Transportation, Viwang, South Korea
j-oh@ut.ac.kr
[3] Chungbuk National University, Cheongju, Chungbuk, South Korea
jjung@chungbuk.ac.kr

Abstract. The Permeable block pavement is one of the representative Low Impact Development (LID) techniques for innovative water management based on systems that can infiltrate usually over the surface. The material of each layer in permeable block pavements is often comprised of a granular material with very high porosity not like conventional pavements. However, unlike conventional pavements, researches on structural characteristics of permeable block pavement is insufficient yet. Therefore, in this study, as part of study to understand the structural characteristics of permeable block pavement and open graded aggregate that is used as base material of permeable block pavement, plate load test was carried out on two different pavement sites, site M and N. Site M is aggregate base sites and site N is permeable block pavement site. Based on the relationship between load and settlement that is obtained from the test, coefficient of subgrade reaction (k) was calculated. Through the relationship between load-settlement and k, structural characteristics were evaluated according to material differences of the pavements.

Keywords: Permeable block pavement · Coefficient of subgrade reaction
Plate load test · Low impact development

1 Introduction

1.1 Overview

Low impact development (LID) techniques are emerging as effective measures globally to mitigate water cycle distortion caused by climate change and urbanization (Han 2011). In Korea, the GI and LID demonstration center was opened in Pusan National University in 2016, and the demonstration and research on LID technology is being actively carried out in line with this trend. As a representative LID technique, permeable pavements that allow infiltration into the pavement using materials with high porosity, not like traditional pavement systems that do not allow infiltration, is

Z. Hossain et al. (eds.), *Solving Pavement and Construction Materials Problems with Innovative and Cutting-edge Technologies*, Sustainable Civil Infrastructures,
https://doi.org/10.1007/978-3-319-95792-0_6

increasingly constructed and used all over the world nowadays (Scholz and Grabowiecki 2007). Especially, permeable block pavement, which is one type of permeable pavements, is widely used for its ability to reduce runoff and water pollution, and also for its structural stability, excellent constructability, and ability to reduce heat island (Smith 2011). Despite these points, there is a lack of research on the structural characteristics of permeable block pavement even though it uses different materials compared to conventional pavement. Therefore, in this study, as part of understanding the structural properties of permeable block pavement which consists of multi-layered system, plate load test was implemented on the aggregate base site and permeable block pavement site installed at the GI and LID center at Pusan National University. Load-settlement relationships and the coefficient of subgrade reaction (k) values of each site were obtained from the test results. Through the relationships and k values, structural characteristics of the pavements were evaluated.

1.2 Objectives

The material being used in the permeable pavement is different from the material being used in the conventional pavement. In the conventional pavement, dense graded aggregate whose gradation is well distributed is normally used as a base material. In the permeable pavement, however, for the permeability, open graded aggregate which has high porosity due to poor gradation is used as the base material. Also, permeable block that is used as the surface layer has high porosity than the conventional surface materials. Therefore, the overall structural characteristics of permeable block pavement could be different from the conventional pavement. In the previous study, structural ability of pavement layers and geogrid reinforced flexible pavements were evaluated through plate load test (Virgil and Godwin 1995; McCartney et al. 2013). Objectives of this study is to examine structural characteristics of permeable pavement and its consisting layers using k values which is obtained from plate load test and pressure-settlement curve.

2 Plate Load Test and Test Location

2.1 Plate Load Test and Coefficient of Subgrade Reaction (k)

Plate load test is widely used test to investigate the strength, deformation, and support characteristics of earthworks and foundation, as well as pavement from the relationship between the size of the load and the settlement amount by applying a load to the ground through the rigid loading plate. The slope of the line connecting the origin and the pressure corresponding to the limit settlement amount is defined as coefficient of subgrade reaction (k) in the pressure-settlement curve obtained from the test result. The value of limit settlement depends on the type of pavement. In the case of block pavement, the load distribution and transmission through the surface layer and base layer is similar with that of flexible pavement (Swan and Smith 2009). In Korean standard (KS F2310 2015), the limit settlement of flexible pavement is 2.5 mm. Therefore, 2.5 mm is used as the limit settlement to calculate k. This k value is used to estimate the elastic modulus of the ground, elastic settlement of the foundation and to design road pavement. Plate load test was done according to KS F2310 (2015) using 30 cm diameter plate.

The advantage of the plate loading test is that the composite bearing characteristics of the soil under test can be considered. In the previous study of Davarifard and Tafreshi (2015), plate load test was done at multi-layered geocell reinforced bed considering embedment depth ratio. Virgil and Godwin (1995) evaluated pavement layer moduli through implementing plate load test on a number of pavement layers in the field. McCartney et al. (2013) evaluated performance of reinforcement in the flexible pavement using plate load test. Since permeable pavement system is multi layered composite soil system, *k* value which can be obtained from plate load test is appropriate measures to evaluate structural ability of permeable block pavement system since they can consider these characteristics. Therefore, in this study, *k* value is adopted to evaluate the structural ability of pavements.

2.2 Test Location

The test was conducted at the surface of open graded aggregate base site (site M) and the surface of permeable block pavement site (site N), which were installed at the GI and LID Center (Figs. 1 and 2).

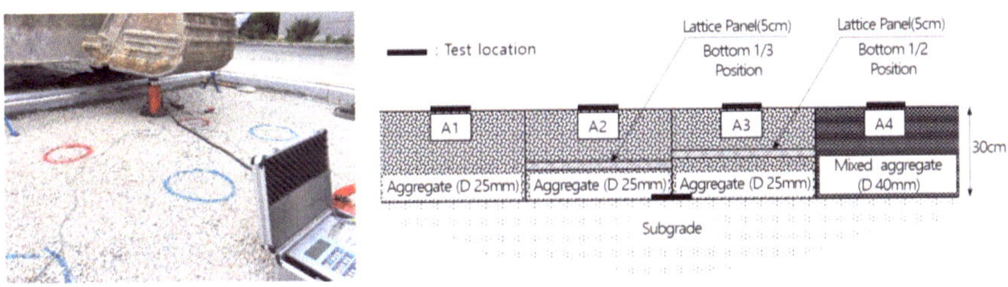

(a) Plate load test overview **(b) Cross section and test locations**

Fig. 1. Site M—aggregate base site

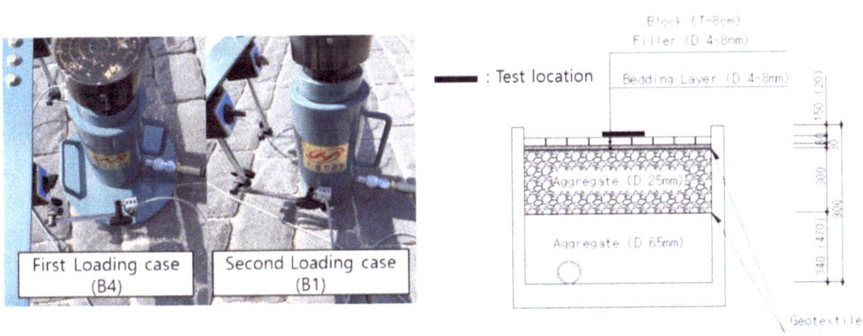

(a) Plate load test overview **(b) Cross section and test location (Lee, 2017)**

Fig. 2. Site N—permeable block pavement site

Site M, aggregate base site, has four different kinds of base layer sections whose dimension is 4 m by 4 m each (Fig. 1b). Two kinds of aggregate consist of the site. One is open graded aggregate whose gradation is poorly distributed with high porosity for permeability, and the other is mixed aggregate whose gradation is well distributed like conventional pavement. Section A1 is composed of D 25 mm open graded aggregate. D 25 mm open graded aggregate is normally used permeable pavement base material which has similar gradation with ASTM No. 57. ASTM No. 57 is recommended as permeable pavement base material gradation according to Smith (2011). Section A2 and A3 is also composed of D 25 mm open graded aggregate, but it has geogrid reinforcement which is named Lattice panel at bottom 1/3 position and 1/2 position each inside the base material. Section A4 is composed of D 40 mm mixed aggregate. D 40 mm mixed aggregate is typical base material which is used in conventional pavement. Figure 1a shows plate load test overview and Fig. 1b shows the cross sections, material compositions and test locations of site M.

Site N, a whole permeable block pavement system whose dimension is 2.3 m by 11 m, is composed of different materials (Fig. 2b). Generally, in the permeable block pavement, the surface layer is composed of permeable block, joint filler, and bedding layer and the base layer and subbase layer are composed of open graded aggregate for permeability. Site N is constructed in compliance with such requirements. Surface layer of site N consists of permeable block, D 4 ~ 8 mm joint filling aggregate and bedding aggregate. Base layer consists of D 25 mm open graded aggregate which is the same material as the one used in site M—section A1. Subbase layer consists of D 65 mm aggregate. Figure 2a shows plate load test overview and Fig. 2b shows the cross sections, material compositions and test location of site N. In the case of Site N, plate load test was conducted in two different loading cases (Fig. 2a) to check structural characteristic difference according to loading area through the block. Because the surface of block pavement is not continuum like asphalt pavement and the plate size is limited, the number of blocks covered by the plate could be different depending on its laid position and size. In the first loading case (B4), the load was applied on the surface through 30 cm plate covering four blocks. In the second loading case (B1), the load was applied directly on one block without plate.

3 Test Results

The results of plate load tests performed at each site are shown in Fig. 3 and Table 1 respectively. Figure 3 is a graph that shows the relationship between pressure and settlement. Table 1 shows calculated *k* values corresponding to a settlement of 2.5 mm.

In site M, A4 section which consists of D 40 mm mixed aggregate, showed bigger *k* compared to A1, A2, A3 which consists of D 25 mm open graded aggregate. This shows that conventional dense graded aggregate has higher structural ability than open graded aggregate that is used in base layer of permeable block pavement. Comparing the results of the section A1, A2, and A3 which consists of same open graded aggregate (Table 1), Lattice panel reinforcement increased *k* value of the base layer by 27% regardless of its installation position. It shows that as long as the reinforcement is installed in a proper position, subtle location or level change of it does not affect much

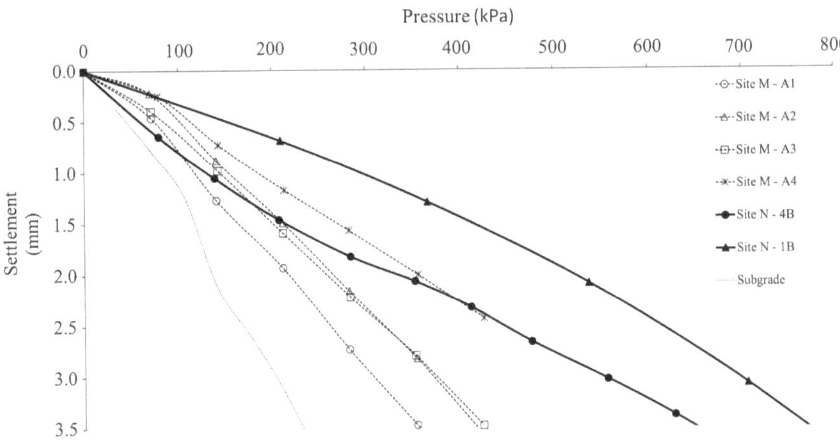

Fig. 3. Plate load test results

Table 1. Coefficient of subgrade reaction (*k*) at site M and site B

Test location		*k* (kN/m^3)
Site M aggregate base site	Surface of subgrade (S)	68,015
	Surface of open graded aggregate (D 25 mm) base layer without reinforcement (A1)	104,000
	Surface of open graded aggregate (D 25 mm) base layer with Lattice panel reinforcement (bottom 1/3 position) (A2)	132,000
	Surface of open graded aggregate (D 25 mm) base layer with Lattice panel reinforcement (bottom 1/2 position) (A3)	132,000
	Surface of mixed aggregate (D 40 mm) base layer (A4)	174,000
Site N permeable block pavement system site	Surface of permeable block pavement system through four blocks (4B)	178,000
	Surface of permeable block pavement system through one block (1B)	244,800

in the structural ability of the base layer. Also, it is considered that the application of this reinforcement can be effective on open graded aggregate base layer to compensate low structural ability due to high porosity.

In site N, comparing the results *k* by loading case, 1B which has smaller loaded area showed higher *k* than 4B, regardless of loaded location in M. It is considered that the decrease of the influence range of stress in the base and subbase due to decrease of loaded area made the settlement smaller. Also, in Fig. 3, compared to N-1B, slope of pressure-settlement curve of N-4B gradually became gentle as the load increased, which seemed to be due to the interlocking effect of the block pavement. The interlocking effect is an effect of improving the structural performance by interlocking between blocks and bedding aggregate and joint filler. However, in N-1B, slope became steeper as the load increased. When the load in only loaded just in one block like N-1B, interlocking effect did not appear.

Comparing site M and N by *k*, site N showed higher *k* overall. Permeable block surface layer which consists of bedding sand, joint filler and permeable block made structural ability higher. As stated above, interlocking effect of those materials made structural ability higher than that of intact base layer without surface layer.

4 Conclusion

By implementing the plate load test on the aggregate base site M and permeable block pavement site N, structural and bearing characteristics can be successfully evaluated and compared based on *k* values. As a conclusion, following facts can be obtained by comparing *k* values.

1. Comparing *k* values, open graded aggregate base which is used in the permeable block pavement base, has lower structural ability than mixed aggregate base which is conventional aggregate base. It is considered that the high porosity of the open aggregate influenced the decrease of structural ability. However, the decreased structural ability could be supplemented by geogrid reinforcement. In this study, geogrid reinforcement increased k value of the base layer about 27%.
2. When the plate load test is implemented on the surface of permeable block pavement, loaded area could be a major factor that caused *k* value change. When loaded area was smaller, settlement became higher than that of bigger loaded area. It is considered that the decrease of the influence range of stress in the base and subbase due to decrease of loaded area made the settlement smaller. Also, in Fig. 3, compared to N-1B, slope of pressure-settlement curve of N-4B gradually became gentle as the load increased, which seemed to be due to the interlocking effect of the block pavement. When the load in only loaded just in one block like N-1B, interlocking effect did not appear.
3. Permeable block surface layer brought more structural ability to the pavement system. Moreover, as the load applied on the surface of the pavement increased, structural performance was increased. The reason seemed to be interlocking effect of blocks, bedding aggregate and joint filler.

Also, depending on the laid position of plate, the number of blocks that is loaded by plate can be changed. This variable might affect *k* value. Further study should be done to consider this variable for more appropriate evaluation.

Acknowledgements. This research was supported by Korea Agency for Infrastructure Technology Advancement (KAIA) grant funded by the Ministry of Land, Infrastructure and Transport (Grant 18CTAP-C132363-02), and Basic Science Research Program through the National Research Foundation of Korea (NRF) funded by the Ministry of Education (No. NRF-2017R1D1A3B03034563).

References

Davarifard, S., Tafreshi, S.M.: Plate load tests of multi-layered geocell reinforced bed considering embedment depth of footing. Procedia Earth Planet. Sci. **15**, 105–110 (2015)

Han, W.S.: US low impact development applications and implications for urban rainwater management improvements. Homel. Policy Br. **344**, 1–6 (2011)

Korea Industrial Standards Commission: Standard Test Method for Plate Load Test on Soils for Road. KS F2310 (2015)

Lee, E., Kim, G., Ahn, J., Shin, H.-S.: Surface infiltration rate of permeable block pavements depending on the size of filling materials. J. Korean Soc. Hazard Mitig. **17**(1), 227–233 (Feb 2017)

McCartney, J., Brady, R., Wood, C., Curry, B.: Evaluation of geosynthetic-reinforced flexible pavements using static plate load tests (2013)

Scholz, M., Grabowiecki, P.: Review of permeable pavement systems. Build. Environ. **42**(11), 3830–3836 (2007)

Smith, D.R.: Permeable Interlocking Concrete Pavements. Interlocking Concrete Pavement Institute (ICPI), Herndon, VA (2011)

Swan, D.J., Smith, D.R.: Development of the permeable design pro permeable interlocking concrete pavement design system. In: 9th International Conference on Concrete Block Paving, Argentina, pp. 18–21 (2009)

Virgil, W., Godwin, H.: Evaluation of pavement layer moduli using field plate bearing load test. Transp. Res. Rec. (1995)

Slope Stability of Tailing Dam Under Seismic Excitation

Pragyan Pradatta Sahoo(✉), Sanjay Kumar Shukla, and Alireza Mohyeddin

Discipline of Civil and Environmental Engineering, School of Engineering, Edith Cowan University, Joondalup, Perth, Australia
pragyans@our.ecu.edu.au, {s.shukla,mohyeddin}@ecu.edu.au

Abstract. The slope stability analysis of the tailing dams or mine dumps becomes a more challenging task for engineers when they are subjected to a combination of static and dynamic loadings. Several tailing dam failures have been observed in the past worldwide. The failure of tailing dam not only releases the waste deposit to the surrounding but also yields the property damage and cause environmental threat to the locality. The waste deposit may liquefy under the earthquake with an increase of the horizontal load on the saturated dam tailing. Hence failure of the dam under liquefaction may lead to catastrophe. In 2015, Samarco tailing dam has failed under the earthquake condition. When the dynamic loading associates with horizontal seismic load along with vertical load, it may reduce the stability of slope leading to failure. A two-dimensional numerical attempt has been made by utilising the commercial software Geostudio, and the same is presented in this paper. Initially stability of the slope is evaluated by considering the finite element stress analysis. A staged pseudo-static analysis has been presented to the study the effect of factor of safety of tailing dam under the action of both horizontal and vertical seismic loads.

Keywords: Horizontal seismic coefficient · Factor of safety · Tailing dam Pseudo-static analysis · Vertical seismic coefficient

1 Introduction

Failure of several tailing dams under seismic action is one of the common incident observed in the past. The failure of these tailing dams becomes hazardous to the environment and the mankind due to discharge of toxic waste deposits to the surrounding. Hence, utmost safety measures need to be taken care by the geotechnical engineers for the design of tailing dams under earthquake loadings to minimise the damage of the tailing dams. From the earlier research, it is observed that possible failure is due to the loss of freeboard followed by subsequent overtopping or settlement of the crest of the dam due to strain generated due to cyclic loading in the earthquake process (Makdisi and Seed 1977). Clough and Chopra (1965) investigated the dynamic response of the dam for a planestrain condition considering the linearly elastic,

Z. Hossain et al. (eds.), *Solving Pavement and Construction Materials Problems with Innovative and Cutting-edge Technologies*, Sustainable Civil Infrastructures,
https://doi.org/10.1007/978-3-319-95792-0_7

homogenous and isotropic soil system. In most of the research, behaviour of the dam has been studied by considering the horizontal acceleration of the earthquake without giving due weightage to the vertical earthquake acceleration. However, the effect of vertical earthquake acceleration plays a vital role for the failure of any earth structure under seismic condition (Shukha and Baker 2008; Stewart et al. 1994).

Numerous failure of tailing dams has been reported in the past. The catastrophic failure of the Fundao tailing dam near Bento Rodrigue village, Brazil has caused a huge economical loss. Therefore, an assessment of performance is required in the design of geotechnical structures under seismic condition. This can be achieved by expressing the factor of safety against the slope failures or deformations. Here, the performance of the dam has been studied under staged pseudo-static analysis for various combinations of horizontal as well as vertical earthquake loadings to evaluate the stability under seismic condition.

Currently, several geotechnical commercial software platforms are available for the seismic analysis of earthen structures such as dams, slopes and embankments. Finite element based methods such as PLAXIS, ABAQUS or FLAC, including SLOPE/W, a limit equilibrium approach are used widely by practising engineers or the researchers (Chakrobarty and Choudhry 2011; Jakka et al. 2011) for the analysis of dams both in static as well as in dynamic cases. In the present paper, both static and pseudo-static analyses of a typical actual section of Fundao dam located in Brazil have been made. The geotechnical properties of the soil and static strength parameters has been chosen from the dam reports (Samarco 2016) for the purpose of analysis. The objective of the analysis is to study the impact of combined seismic action on the factor of safety of the dam under staged pseudo-static analysis. Combined seismic action on the slope may lead to failure for the small magnitude of the earthquake which has been demonstrated in the current study. The analysis of the tailing dam model is mainly performed with the limit equilibrium method by using SLOPE/W.

2 Problem Specification and Analysis

Both static and pseudo-static case analyses are performed for the left abutment of the tailing dam section. The height of the tailing dam section above the existing ground level is considered 56 m having downstream slope 6.2H:1V. The dam consists of slimes and loose sand tailings above the ground level. Additionally, the dam is supported by the thin compacted shell of sand tailings of width approximately 20 m. The pore water condition of the section was based on cone penetration tests and was conducted by using an integrated electronic piezocone penetrometer and data acquisition system to record pore pressure at various locations. These data have been used in for the stability analysis of the tailing dam. Table 1 presents the properties of various components of the tailing dam reported by the Fundao dam committee (Samarco 2016). The engineering properties of the soil are based on the field data and laboratory data such as direct shear test were used to determine the effective cohesion and friction angle of the tailing dam materials. Slope stability analysis was undertaken for the failure conditions in November 2015 model. Figure 1 shows the model geometry of the left abutment, being analysed in 2D plane-strain condition on SLOPE/W (2007).

Table 1. Effective stress parameters for the dam section

Types	Material	Unit weight (kN/m^3)	Effective cohesion (kPa)	Effective angle of internal friction (degrees)
Tailings	Compacted tailings	22	5	35
	Loose sand tailings	22	0	33
	Slimes	22	0	28
Foundation	Weathered phyllite	22	40	32

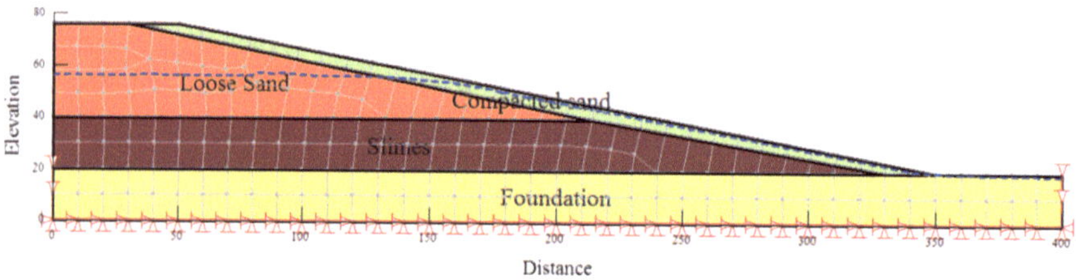

Fig. 1. Details of the left abutment of Fundao dam

With the different components, the tailing dam has modelled in geotechnical software Geostudio with the following steps:

(1) SIGMA/W: In situ static stress has been computed.
(2) SLOPE/W-static: At this stage, the factor of safety the tailing dam is calculated under the static condition.
(3) SLOPE/W-pseudo-static: Seismic load has been applied to the model.

Initially, in situ static stress that considers only the gravity loading of the dam has been carried out by using SIGMA/W and the factor of safety of the dam under such condition has been evaluated. For the entire analysis in SLOPE/W, Mohr-Coulomb plasticity soil model has been used to characterize the behaviour of the soil material. Factor of the safety of the dam has been calculated by using Morgenstern-Price limit equilibrium method. For the pseudo-static analysis of the tailing dam, the input peak horizontal acceleration of 0.15 g of earthquake magnitude 2.6 has been used for the analysis. The earthquake acceleration is applied in both horizontal and vertical directions of the dam considering the vertical acceleration is being one-half of the horizontal acceleration (IS 1893 2002). The critical factor of safety of the dam is computed by considering the two different combinations of earthquake accelerations:

(1) Horizontal acceleration ($k_n g$) and positive vertical acceleration ($k_u g$) (↓)
(2) Horizontal acceleration ($k_n g$) and negative vertical acceleration ($k_u g$) (↑)

3 Results and Discussion

In the analysis, an effort has been made to observe the effect of combined seismic action for the failure of the dam by varying the both pseudo-static acceleration with factor of safety. Initially, the stability factor is evaluated for the gravity loading condition. The static factor of safety obtained from finite element stress method has been compared with the factor of safety under the application of seismic loads. Figure 2 shows the factor of safety of the dam section under static condition and is found to be 1.6. This signifies that the slope is stable under in situ condition.

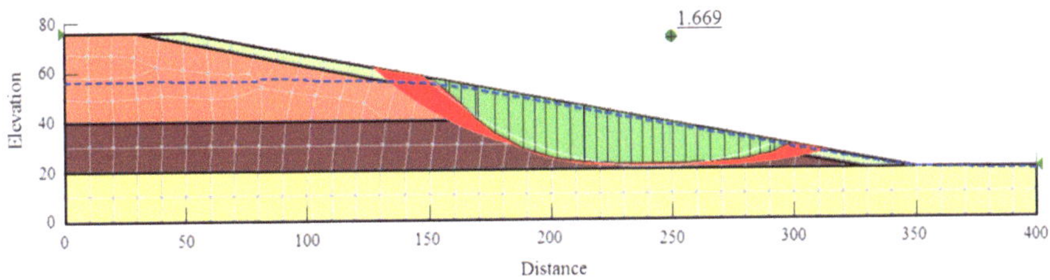

Fig. 2. Static slope stability factor under in situ condition

Figure 3 shows the factor of safety for the tailing dam section obtained under the combined seismic actions. Both horizontal ($k_h g$) and vertical ($k_v g$) seismic accelerations are applied in the soil model, considering the direction of action of vertical seismic acceleration being downward (↓). For $k_h = 0.15$, the factor of safety is found to be 0.95. This indicates that the tailing dam is apparently unstable under earthquake condition when the magnitude of horizontal seismic coefficient increases. During earthquake, the vertical component of the ground motion may either act vertically upward or downward directions of the tailing dam. Therefore, an effort has also been made to evaluate factor of safety of the dam considering the upward direction (↑) of the vertical acceleration of the ground motion. Figure 4 shows that the factor of safety of the slope with upward vertical seismic acceleration is found to be 0.862 which is more critical when compared to case presented in Fig. 3. This indicates that vertical seismic acceleration may reduce the stability of the slope considering its direction of loading.

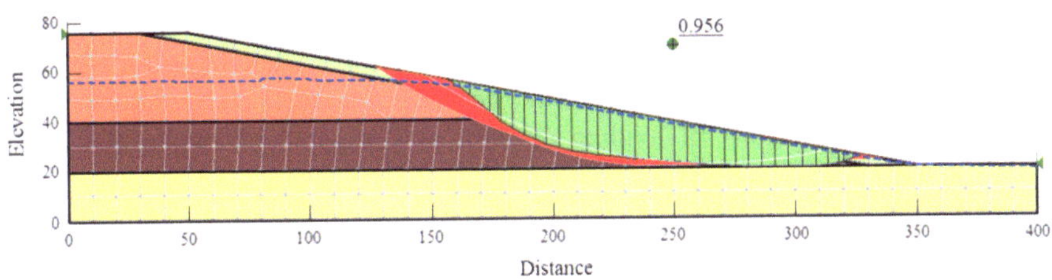

Fig. 3. Static slope stability factor under seismic condition with vertical acceleration in the downward direction (↓)

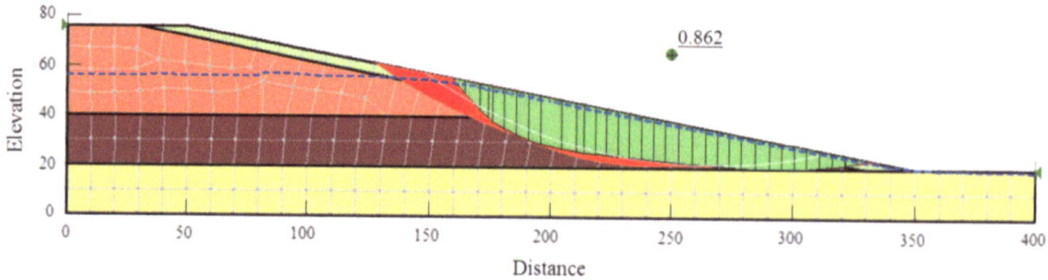

Fig. 4. Static slope stability factor under seismic condition with vertical acceleration in the upward direction (↑)

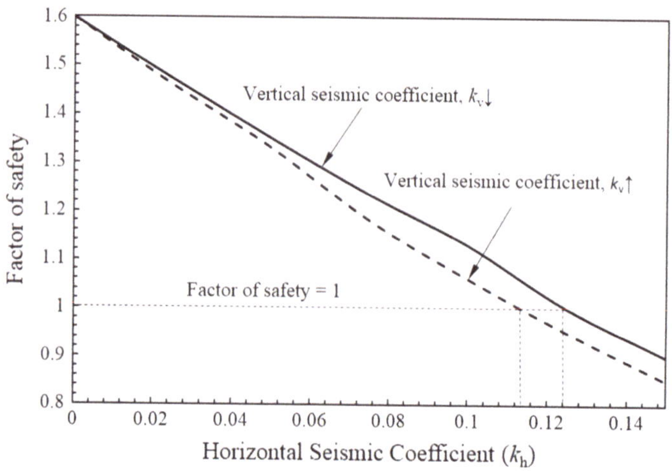

Fig. 5. Variation of factor of safety with horizontal seismic coefficient

Figure 5 shows the variation of the pseudo-static factor of safety with different values of horizontal seismic coefficient (k_h) taking account of the application of vertical seismic coefficient (k_v) on the tailing dam. The factor of safety of the dam has been obtained by considering the staged pseudo-static combined seismic loading condition with upward and downward application of vertical seismic coefficients. From Fig. 5, it is noticed that factor of safety of the tailing dam reduces almost linearly with increase of horizontal seismic coefficient for both the curves. It may be further noted that the factor of safety is found to be unity when the horizontal seismic coefficient is 0.124 (k_v ↓) and 0.112 (k_v ↑) considering both directions of loading of vertical seismic coefficients. Under such conditions, these horizontal seismic coefficients are termed as yield seismic coefficients (Newmark 1965). Thus, yield acceleration for the dam section can be computed from Fig. 5 considering the effect of both horizontal and vertical seismic accelerations of the earthquake. From Fig. 5, the critical yield acceleration is computed as 0.112 g. Hence, it may be further inferred that the yield seismic coefficient has lower value when vertical seismic load acts against the direction of gravity of the tailing dam. Hence, there is a high probability of the dam deformation in the shortest duration of the seismic loading condition considering critical condition of combined seismic loadings.

In the past, most of the researchers have tried to compute the displacement to assess post-earthquake effects on the dams based on Newmark (1965) sliding block model. One of the empirical expression of permanent displacement of the dam caused by the ground motion is given by Ambraseys and Menu (1988):

$$\log u = 0.90 + \log\left[\left(1 - \frac{a_y}{a_{\max}}\right)\left(\frac{a_y}{a_{\max}}\right)^{-1.09}\right] \quad (1)$$

where u is the displacement in cm.

$a_{\max}$ is peak ground acceleration.

a_y is the yield acceleration.

Using the above expression and values of $a_{\max} = 0.15g$ and $a_y = 0.112g$, the displacement of the dam section is found to be 13 cm. However, this value gives more conservative value towards the displacement as the empirical formula is mostly based on the horizontal ground acceleration without giving weightage to vertical seismic effect. Hence, it is encouraged to calculate the displacement either by use of numerical modelling software or to establish a new analytical formulation to evaluate displacement of the tailing dam under combined seismic actions.

4 Conclusions

The effect of vertical acceleration of the ground motion on the dam has not been studied extensively in the past. In the present study, an attempt has been made to study the failure of the dam under the effect of both horizontal seismic acceleration and vertical seismic acceleration during earthquake condition. Based on the basis of results and discussion, the following conclusions can be drawn:

(1) Factor of safety of the tailing dam is found to be maximum under static condition and more stable when compared to the seismic condition.
(2) The critical factor of safety of the dam section is found to be 0.862 under combined effect of seismic coefficients, with vertical seismic load being vertically upward direction.
(3) The yield acceleration of the slope under staged pseudo-static analysis is found to be 0.112 g.

It can be finally concluded that effect of vertical seismic coefficient is need to be included in the seismic design of the dam analysis.

References

Ambraseys, N.N., Menu, J.M.: Earthquake-induced ground displacements. Earthq. Eng. Struct. Dyn. **16**(7), 985–1006 (1988)

Chakraborty, D., Choudhury, D.: Seismic behaviour of tailing dam using FLAC. In: Geo-Frontiers 2011: Advances in Geotechnical Engineering, pp. 3138–314 (2011)

Clough, R.W., Chopra, A.K.: Earthquake stress analysis in earth dams. J. Eng. Mech. Div. **92**(2), 197–212 (1965)

Samarco: http://fundaoinvestigation.com/the-panel-report/(2016)

IS: 1893: Indian Standards Criteria for Earthquake Resistant Design of Structure. Part-1. Bureau of Indian standards (BIS), New Delhi, India (2002)

Jakka, R.S., Ramana, G.V., Datta, M.: Seismic slope stability of embankments constructed with pond ash. Geotech. Geol. Eng. **29**(5), 821–835 (2011)

Makdisi, F.I., Seed, H.B.: Simplified procedure for estimating dam and embankment earthquake-induced deformations. In: ASAE Publication No. 4–77. Proceedings of the National Symposium on Soil Erosion and Sediment by Water, Chicago, Illinois, 12–13 December 1977

Newmark, N.: Effects of earthquakes on dams and embankments. Geotechnique **15**(2), 139–160 (1965)

Shukha, R., Baker, R.: Design implications of the vertical pseudo-static coefficient in slope analysis. Comput. Geotech. **35**(1), 86–96 (2008)

SLOPE/W: GeoStudio Student Edition [Computer Software]. GEOSLOPE International Ltd, Calgary, AB, Canada (2007)

Stewart, J.P., Bray, J.D., Seed, R.B., Sitar N.: Preliminary Report on the Principal Geotechnical Aspects of January 17, Northridge Earthquake. Report No. UCB/EERC-94/08, Earthquake Engineering Research centre, University of California, Berkeley (1994)

Use of the Shear Box Compactor for Porous Asphalt Mix Property Assessment

Irina Holleran[1(✉)], Douglas J. Wilson[1], Glynn Holleran[1], and Lubinda F. Walubita[2]

[1] Department of Civil and Environmental Engineering, The University of Auckland, Auckland, New Zealand
imot002@aucklanduni.ac.nz

[2] Texas A&M Transportation Institute (TTI), The Texas A&M University System, College Station, TX, USA

Abstract. The most commonly used compaction methodologies for specimen preparation for the assessment of porous asphalt (PA) mix properties are gyratory and Marshall. Even though several individual blocks prepared by one of these test methods are usually considered to be replicates, the heating and handling histories are never identical as these are individual specimens prepared at different times. In contrast, the Shear Box Compaction (SBC) technique allows fabrication of a single slab, at a given time, which is then used to extract multiple specimens that can be considered as true replicates. The SBC can offer significant advantages in preparation of PA specimens for various tests typically performed on gyratory or Marshall specimens. The recently released ASTM D7981-15 test method on SBC of asphalt mixes is written around compaction of dense-graded (DG) mixes, not PA mixes, even though the specific mix type is not stated in the test method. Furthermore, there is a very limited literature on the SBC, with most of the published work focusing on DG mixes. This paper aims to address this gap. In this study, the SBC was used for PA mix compaction with a nominal maximum aggregate size of 10 mm using two different types on aggregates and four bituminous binders. Slab preparation and specimen extraction methodology developed were discussed in details. Variation of the air voids (AV) between the replicate specimens extracted from one slab were within ±0.5% from the average AV value regardless of the aggregate or binder type, or slab height. The total variation of AV in all the 64 extracted specimens tested was 1.9%. This gives the confidence that performance testing such as Dynamic Modulus testing, which requires very stringent control of specimen dimensions, uniformity, and AV consistency among the replicate specimens, can be performed reliably on PA. Also, some recommendations were provided on the changes needed in the ASTM D7981 for PA compaction using SBC.

1 Introduction

Gyratory and Marshall are common compaction modes used for specimen preparation of porous asphalt (PA) in the laboratory. Gyratory compaction is typically used in the US methodology for PA mix designs and mix property assessment (Kandhal 2002) as this way of compaction better replicates field compaction compared to Marshall, which

Z. Hossain et al. (eds.), *Solving Pavement and Construction Materials Problems with Innovative and Cutting-edge Technologies*, Sustainable Civil Infrastructures,
https://doi.org/10.1007/978-3-319-95792-0_8

is an impact type method. Nevertheless, regardless of the gyratory or Marshall compaction method selected, individual specimens fabricated using these methodologies are never identical even though these specimens are from the same laboratory prepared or production based asphalt mixes. In contrast, the Shear Box Compactor (SBC) is capable of producing a slab suitable for extraction of multiple specimens, which make this method less time consuming and more accurate due to the elimination of variables during mix handling and compaction. Specimens extracted from the same SBC slab can be considered as true replicates as they were compacted under identical conditions. Gabrawy (2000) suggested that SBC provides much better representation of field compaction compared to gyratory compaction methodology, as it closely replicates the action of a field roller and is capable of compacting large-scale specimens.

There is a limited number of tests typically conducted for the assessment of PA mixes. This includes Cantabro, Tensile Strength Ratio (TSR), permeability and Indirect Tensile Strength (ITS) tests. However, the majority of tests available for the assessment of PA are not directly related to real field conditions, which is an issue for accurate field performance prediction. This is especially true for the Cantabro test that is performed using a rotating Los Angeles Abrasion steel drum under stress conditions that do not simulate the stress generated by traffic movement (Hamzah et al. 2012). So, it can be only used for comparative purposes for selection of the optimum binder content for any given mix during the mix design process. It has been reported that Dynamic Modulus (DM) testing can provide important information in relation to PA mix behavior. This includes understanding the effect of reclaimed asphalt pavement materials (RAP) in PA mixes compared to virgin PA mixes (Goh and You 2012). Also, the effect of different binder types on the PA mix properties can be directly evaluated. In New Zealand (NZ), the highly diluted (to reduce high cost) epoxy-modified binder was introduced as a solution for increasing the durability of PA mixes. This binder is a 25% dilution of 30% epoxy polymer mixed with selected bitumen and the standard 80/100 NZ sealing grade bitumen. This means the actual epoxy content is around 7–8% in the final binder. Based on Cantabro test results, a life span of 44 years was predicted for PA made with this highly diluted epoxy-modified binder. To evaluate this claim, the behavior of PA mixes made with polymer-modified, unmodified and epoxy modified binders were assessed by Holleran et al. (2017) using multiple testing methodologies including DM. It was found that the PA mixes made with polymer-modified binder were outperforming the PA mixes made with the diluted epoxy-modified binder. The DM results showed that this epoxy binder was not curing in the PA mix at ambient temperature for a prolonged period of time. If there is no curing, there are no benefits in relation to slowing the binder aging. The DM testing also indicated that the PA mix made with the diluted epoxy-modified binder may have issues with damage under early traffic and densification in the long term. The diluted epoxy based PA mix had the lowest modulus across the entire range of service temperatures and frequencies compared to the PA mixes made with polymer-modified and unmodified binders. The PA mix behavior was consistent with rheological testing performed on the binders. No evidence was found that the PA made with the diluted epoxy-modified will provide a longer life compared to the polymer-modified PA. It is clear that mechanical properties such as dynamic modulus are useful in assessing the structural performance of mixes across the whole service range of temperature and traffic loading.

The DM test was originally developed for characterization of dense-graded (DG) mixes, not PA. PA is often used as a surfacing mix in many countries with no structural value assigned to this layer, however some countries do use it as a structural layer (Cooley et al. 2009). DM and phase angle are fundamental mechanical properties of asphalt mixes, regardless if it is DG or PA. By doing DM on PA mixes, a better understanding of PA mix fundamental properties such as mix behavior under different traffic and temperature conditions can be achieved resulting in a better prediction of field performance.

Sample preparation for DM testing is one of the critical steps to get representative results. AASHTO standards for specimen compaction (AASHTO 2009) and DM testing (AASHTO 2013) require at least three gyratory compacted specimens with very stringent control of mix homogeneity and specimen dimensions.

SBC compaction offers significant advantages compared to gyratory compacted specimens for DM testing. From one single SBC slab, up to four 100 mm diameter and 150 mm in height specimens can be extracted. However, there is no standard methodology for compaction of PA mixes using SBC currently available. It took around 21 years from the first SBC prototype developed in Australia in 1994 (Gabrawy 2000) to the release of the new ASTM D7981 (ASTM International 2015) test method on compaction of asphalt mixes using SBC. This standard is tailored for compaction of DG mixes, not PA mixes, even though the type of asphalt mix is not mentioned in the standard. For example, in the ASTM D7981 standard, the height limitation required for a SCB slab is 185 mm maximum. This is a suitable limit for DG mixes, but not for PA, as will be demonstrated in this paper. Typical air voids (AV) in PA specimens are in the range of 20–25%, which is much higher compared to DG mixes, where 7 ± 0.5% AV is required for DM testing. So, SBC PA slab heights can be much higher than the 185 mm limit. This means that the extraction of larger number of specimens is possible from a single slab to conduct the multiple tests traditionally carried out on gyratory compacted specimens.

Slab preparation, and specimen extraction methodology have a significant impact on the quality of PA specimens. PA specimens have tendency to lose aggregate particles during specimen handling, such as coring and sawing. DG mixes on the other hand, have a much denser structure, so the aggregate particles are not as easily dislodged from the mix matrix. Loss of aggregate particles from PA specimens will affect the AV result calculation as it is based on specimen weight, volume, and maximum theoretical specific gravity as per ASTM D3203/D3203M (ASTM International 2011a) test method. Furthermore, the way PA specimens are dried after trimming will have an effect on AV results as it affects the final weight of the specimens. Water from coring and sawing of specimens from a SBC slab permeate through the PA specimens and can get trapped within the specimen due to its porous interconnected structure. All these factors combined, make PA sample preparation more challenging in relation to achieving relatively close AV content between the replicates compared to DG. This is especially true for specimens for DM testing which requires AV to be within 0.5% from the average value. Zhang et al. (2013) conducted DM testing on DG and PA mixes. The AV content in the gyratory compacted specimens for PA was 2% from the average AV to 0.5% for DG specimens.

Overall, there are very limited publications available on the SBC. The previously conducted research on SBC (Gabrawy 2000; Molenaar et al. 2009; Priyadharshini and Krishnan 2014; Qiu et al. 2012) concentrated on compaction and volumetric testing of DG mixes. From the literature reviewed at the time of writing this paper, the authors did not find any publications related to the evaluation of AV variability of the specimens extracted from an SBC slab compacted from PA mixes. Also, no research has been published on SBC slab preparation and specimen extraction from SBC slabs compacted using PA mixes. This paper fills in the gap in knowledge in the area of using SBC on PA mixes. The proposed methodology covers slab preparation, coring, sawing, and drying of PA specimens for DM testing including the AV distribution within a SBC slab and between the slabs made with different PA mixes.

2 Research Motivation and Objectives

In NZ, no inclusion of reclaimed asphalt pavement (RAP) is allowed in PA mixes. In the next 6 years, the level of RAP available from PA sites on Auckland Motorways is projected to increase sharply. To be able to recycle this RAP into new PA mixes, a better understanding of PA fundamental properties is essential. DM is a very useful test for the modulus/stiffness assessment of PA mixes with and without RAP. However, sample preparation for DM testing on PA mixes can be very challenging to be able to achieve low AV variability in the replicates. If the AV in replicates are not strictly controlled, this is likely to have a significant effect on the DM test results. The objective of this research is to develop a SBC specimen preparation methodology suitable for DM testing of PA.

3 Materials

PA mixes with nominal maximum size aggregate size (NMAS) of 10 mm, the most commonly used NMAS in PA in NZ, were used in this study. PA mixes were prepared with four different binder types: (1) performance-graded (PG) polymer-modified binder PG64 Grade V, which meets the requirements of NZ performance-based specification M-1A for hot mix asphalt (NZ Transport Agency 2016); (2) performance-graded (PG) unmodified binder PG64 Grade H (NZ Transport Agency 2016); (3) thermoset-modified binder; and (4) thermoset-modified binder with additive.

Two different types of aggregates were utilized in the PA mixes: (a) greywacke aggregates; (b) predominately melter slag and a small amount of greywacke aggregate (less than 10% of the total coarse aggregate fraction). Information on each mix is shown in Table 1. Greywacke is the most used aggregate for road construction in NZ, including in PA mixes. Melter slag, a by-product from steel production, is primary used on areas where skid resistance enhancement is required.

Table 1. PA mix information

Mix name	Aggregate type	Binder type	Binder content (%)
PA10-G-PG64H	Greywacke	PG64 Grade H	5.2
PA10-G-TM	Greywacke	Thermoset-modified	5.2
PA10-G-PG64V	Greywacke	PG64 Grade V	5.2
PA10-S-TM	Slag with greywacke	Thermoset-modified	5.5
PA10-S-TMA	Slag with greywacke	Thermoset-modified with additive	5.5
PA10-S-PG64V	Slag with greywacke	PG64 Grade V	5.5

4 Experimental Work

4.1 PA Slab Preparation Methodology Development

SBC PReSBOX™ apparatus was utilized for slab preparation. Mix handling and compaction were carried out in accordance with ASTM D7981, apart from height restriction and use of a riffle box as a chute to fill the mold. The slab had the nominal dimensions of 450 mm in length and 150 mm in width, which are fixed as these are internal dimensions of the mold. The slab height restriction to the maximum of 185 mm was not followed on purpose. The use of the loading chute was not utilized due to several factors. Firstly, PA mixes are gap-graded consisting mainly of coarse aggregate fractions, so they do not segregate like DG mixes of the same NMAS during sample loading in the SBC mold. Secondly, this chute restricts asphalt mix amounts that can be loaded in the mold for the compaction greater than 185 mm height slabs. This is because the chute occupies some volume of the mold when inserted. It also makes it hard to see the maximum volume of asphalt mix that can be fed into the mold. And finally, it slows down the mix loading which can result in temperature loss and affects the uniform heating history within the slab.

The SBC mold didn't have any means of heating unlike the gyratory mold, which can be pre-heated in an oven prior to mix loading. The mold can thus act as a heat sink rapidly cooling the mix. The high voids content of the PA also means it cools faster than DG mixes. So, the loading of the PA asphalt mix needs to be relatively rapid otherwise it can result, in non-representative final AV and thus, a heterogeneous structure. The target compaction temperature used for PA slab compaction was 130 ± 5 °C (Standards Australia 1995), which is the typical temperature for PA mix manufacture in NZ for laboratory preparation and plant production.

It has been reported that SBC can be used to determine workability of asphalt mixes (Qiu et al. 2012). But if the mold temperature is a variable due, for example, to changes in the temperature in the lab in winter versus summer, it may give misleading mix workability results for mix comparison. Also, the AV level in PA mixes can vary from one mix to another, which can result in different cooling rates in this mold. The current recommendation of the equipment manufacturer, is to heat the asphalt mix in the oven at the higher than compaction temperatures during the mix conditioning stage to compensate for the temperature of the cold mold. In the view of the authors, these are not optimum handling conditions for asphalt mixes especially for PA. PA mix, which is heated above the mix nominated compaction temperature during the conditioning stage, can generate excessive binder drain down resulting in a non-representative mix. The rate of binder drain down will depend on the NMAS of the PA mix, gradation, and binder type. In the case of warm mixes, regardless of the asphalt mix type, the temperature control of the mix is critical and heating it above the nominated compaction temperature will result in non-representative asphalt mix properties. It is very important to eliminate this variable and be able to heat the mold to the target compaction temperature. Due to the unavailability of a custom-designed and built mold heater during this research work, it was necessary to condition all PA mixes at 140–145 °C in an air-forced oven so as to be able to achieve the compaction temperature of 130 ± 5 °C.

Once the mix was placed in the mold, a preheated metal rod was used to probe the edges of the mold to ensure that there were no air pockets between the mold and the PA mix, see Fig. 1a. While performing this step, a temperature probe was inserted in the mix to record the mix temperature. Then, a preheated wear plate and top platen were placed on the top of the asphalt mix, and the mold was inserted into the machine to start compaction. All SBC slabs were compacted using a vertical stress of 600 kPa and shear angle of 4° to a target AV of 25%. The required air void content in 10 mm NMAS PA mixes in NZ is 20–25% (NZ Transport Agency 2005, 2007).

(a) (b) (c) (d)

(e) (f) (g)

Fig. 1. Slab preparation: **a** preparing the slab for compaction; **b** warming the mold for slab extraction; **c** extraction of slab; **d** warming the metal plates for removal; **e** metal plates removed from the top and bottom of the slab; **f** epoxy resin applied on the top and bottom of the slab; **g** slab in a frame with clamps for storing before coring

When the slab compaction process was completed, the mold was unlocked, pulled out of the machine and left overnight to cool down with the sample still sitting in the mold. Extraction of the cold slab from the mold was carried out by warming the SBC mold with a gas torch as shown in Fig. 1b. Trials were carried out to develop a methodology to minimize the damage to the slabs during extraction from the SBC mold. It was found that complete cooling of the slab is essential—this restricted the preparation of one slab a day. PA slabs could deform, collapse or lose aggregate particles when extracted while hot or warm. The weight of the SBC slab is over 20 kg, which is significantly higher than gyratory and Marshall compacted specimens. PA has substantially greater AV than DG mixes, and its structure highly relies on coarse aggregate interlock and the binder rheology. The PA slabs must be allowed to fully cool in the mold and be supported when handled. The procedure on slab ejection outlined by Qiu et al. (2012), where a slab is ejected from the mold first, and then left to cool down completely, is not suitable for PA mixes. DG mixes have significantly denser structure compared to PA mixes and do not deform under their own weight in these circumstances.

After extraction of the slab from the SBC mold (see Fig. 1c), a gas torch was used to warm up the top and bottom metal plates, which are adhered to the slab (Fig. 1d). A minimal amount of asphalt mix was left on these plates after the plate removal (Fig. 1e). Measurements of slab dimensions and weight were recorded and then approximately 100 g of an epoxy resin (the same epoxy which is typically used for polished stone value shoe preparation) in total was applied to the top and bottom of the slab (Fig. 1f). During the trials on coring of SBC slabs, it was evident that the coring action significantly damaged the top and bottom parts of the slab by dislodging aggregate particles from the slab. By epoxying the top and bottom parts of the slab, the damage from coring was completely eliminated. The amount of material applied was just enough to hold the aggregate particles together without completely filling the AV on the surface of the slab. The epoxied part was removed during the trimming of the specimens to the target height. A frame to support the slab was placed around the slab with clamps tightened to prevent slab deformation under its own weight if unsupported, see Fig. 1g. Plastic film was placed around the slab to prevent the slab adhering to the frame. The slab was left overnight for the epoxy to set before any coring was carried out.

4.2 Coring

Once the epoxy is set, the plastic film around the slab was removed and the slab was placed back to the frame with clamps for coring to extract four 100 mm diameter specimens, as shown in Fig. 2a–c. Keeping the frame around the slab was important as it insured that no slab deformation occurred under the force of the drilling jig. After extraction, the specimens were placed under a fan overnight before trimming. This was because wet PA specimens are significantly more prone to damage, such as aggregate dislodging from the mix matrix. By fan drying the specimens prior to trimming, the damage to the specimens was minimized.

Fig. 2. Extraction of specimens from SBC slab: **a** coring in the frame; **b** extracted specimens with no damage; **c** close up photo of the epoxied surface

4.3 Trimming

Trimming of the specimens to the target height of 150 mm for DM testing was carried out using an automated saw with a pre-set value for the height of the specimen. This saw makes the trimming process very user-friendly and accurate. Each specimen was trimmed in a plastic tube to eliminate or minimize chipping, see Fig. 3a and b. Without the plastic tube, PA specimens would be prone to damage making such specimens unusable for DM testing. After trimming was completed, the specimens were left on the lab bench under a fan for overnight drying.

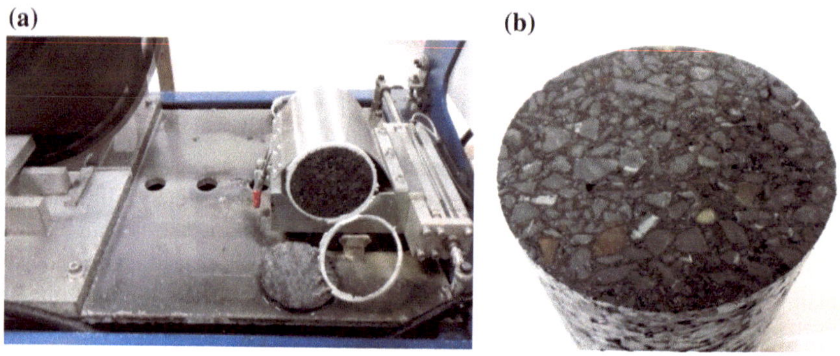

Fig. 3. Specimen trimming to achieve the target height: **a** specimen trimming to the target height of 150 mm; **b** surface of the specimens after trimming

4.4 Drying

Specimens get saturated during the wet sawing and trimming process and moisture can be trapped inside the porous structure. Qiu et al. (2012) reported that DG specimens extracted from SBC slabs were dried with high pressure air and left for 24 h on a lab bench before testing. This methodology was found unsuitable for PA mixes. The high-pressure air can instead dislodge aggregate particles in wet PA specimens, so this drying method was not used in this study. Furthermore, it was not possible to dry the PA specimens completely using fan drying only, even after leaving specimens for a week. Exposing the specimens in air and light for a prolonged period of time while drying is not desirable as it may affect the performance test results due among others to possible short-term oxidative aging. It was clear that apart from utilizing a fan, vacuum drying needed to be applied to PA specimens to completely draw moisture from the specimen internal structure. Vacuum drying was carried out as per ASTM D7227/D7227M (ASTM International 2011b), as shown in Fig. 4a, b, and resulted in completely dried samples allowing representative AV values to be attained. The aggregate type used for PA preparation influenced the drying time. Melter slag used in PA mix has highly vesicular voids throughout the aggregate particles, so moisture could be trapped in these voids. This made the drying process of PA with melter slag more time consuming compared to PA made with greywacke aggregates. Figure 4c and d demonstrate the difference between PA specimens made with these aggregates. Several stages of fan drying and vacuum drying were applied to achieve fully dried specimens for AV testing.

Fig. 4. Vacuum drying of PA specimens: **a** vacuum drying; **b** message displayed when specimens are not fully dry; **c** extracted PA specimens made with melter slag (left) versus greywacke aggregates (right); **d** close up structure of PA made with melter slag

4.5 Volumetric Testing

Sixty-four specimens in total were extracted from 16 slabs, i.e. four specimens per slab. All specimens met the requirements of AASHTO PP 60 (AASHTO 2009) in relation to the average diameter (100–104 mm), standard deviation (SD) of diameter (≤ 0.5 mm), height (147.7–152.5 mm), end flatness (≤ 0.5 mm), and end perpendicularity (≤ 1.0 mm). AV content testing was conducted according to ASTM D3203/D3203M (ASTM International 2011a) for PA. Table 2 provides a list of the test results. The SD was calculated for AV values of the specimens prepared from the same mix. The lowest SD of 0.24 and the highest 0.39 indicate the low variability, which is consistent with the reported data on DG mixes by Qiu et al. (2012) on six different DG mixes compacted using SBC. Specimens of different shape configuration (cylinders and beams) and sizes were extracted and AV measured. The lowest and highest SD was reported, 0.45 and 0.75, respectively. This suggests that in the case of PA mixes, the AV variability between the specimens extracted from SBC slabs compacted from the same mix is at least equal or better compared to DG specimens reported in the literature.

Table 2. AV in PA specimens extracted from SBC

Mix name	Number of specimens	Min AV (%)	Max AV (%)	Average AV (%)	Standard deviation	Coefficient of variation (%)
PA10-G-PG64H	4	21.10	21.69	21.45	0.28	1.32
PA10-G-TM	8	20.60	21.69	21.25	0.39	1.82
PA10-G-PG64V	12	21.69	22.41	22.06	0.24	1.10
PA10-S-TM	4	20.52	21.15	20.88	0.27	1.30
PA10-S-TMA	28	21.12	22.36	21.63	0.29	1.33
PA10-S-PG64V	8	21.33	22.19	21.73	0.34	1.54

The effect of slab height on AV variability of replicates within one slab was investigated, see Table 3. During slab compaction, the slab height was varied from 163 to 230 mm while the AV target of 25% remained constant. Replicates extracted within one slab had an AV variation of within ±0.5% (mostly ±0.3%) from the average AV content regardless of the slab height. This indicates that slab height didn't influence the AV consistency in the replicates. Nevertheless, it is important to have sufficient SBC height to be able to cut the top and bottom parts of the specimen, which are typically more heterogeneous, to achieve the target height of 150 mm for DM testing. SBC slabs compacted to below 183 mm in height were harder to handle during the trimming process. The optimum SBC slab height found for 10 mm NMAS PA mix is 183–190 mm for extraction of DM test specimens. This height makes the trimming relatively easy. If SBC slabs are compacted higher than 190 mm, it can result in unnecessary waste of the asphalt mix.

The total difference between the maximum and minimum AV values in all the 64 PA specimens tested was 1.9%, which is similar to the results of around 2% on DG mixes by Qiu et al. (2012). This indicates that the SBC can produce PA specimens for DM testing with consistent AV content within the same slab and between the slabs regardless of the different binder type and aggregates. Melter slag aggregate used for PA mix manufacture had significantly different shape and texture compared to greywacke aggregates, but it did not influence the final AV results in the extracted specimens.

Table 3. AV in PA specimens extracted from SBC slabs versus parameters recorded by SBC software

Mix	Slab height[a]	Recorded slab AV[b] (%)	Measured avg. AV[c] (%)	AV variation[d] (± %)	Difference in AV[e] (%)
PA10-G-PG64H	183.226	24.9144	21.4159	0.3	3.50
PA10-G-TM	184.464	24.5054	21.0557	0.5	3.45
	184.942	24.7843	21.4516	0.3	3.33
PA10-G-PG64V	185.194	24.9623	22.1649	0.3	2.80
	195.488	24.9859	22.0165	0.1	2.97
	204.776	24.7843	22.0010	0.3	2.78
PA10-S-TM	179.418	24.6274	20.8768	0.3	3.75
PA10-S-TMA	185.196	24.8160	21.6098	0.2	3.21
	184.022	24.8225	21.4405	0.3	3.38
	183.426	24.8186	21.6022	0.4	3.22
	185.205	24.9008	21.8821	0.5	3.02
	230.429	24.7677	21.6275	0.2	3.14
	205.010	24.7551	21.4980	0.3	3.26
	195.615	24.7560	21.7508	0.3	3.01
PA10-S-PG64V	162.636	24.9401	22.0204	0.2	2.92
	185.339	24.7193	21.4488	0.1	3.27

[a]Slab height at the end of the compaction process as recorded by the SBC software
[b]Air void level in the SBC slab at the end of the compaction process as recorded by the SBC software
[c]Average air voids measured in the specimens, 100 mm in diameter and 150 mm in height, extracted from the SBC slabs
[d]Air void variation from the average value for the specimens, 100 mm in diameter and 150 mm in height, extracted from SBC slabs
[e]Difference in the air voids between the recorded slab air voids at the end of slab compaction as recorded by the SBC software and the average measured air voids in the extracted specimens

The average AV content measured on the extracted specimens was 3–4% lower compared to the AV values recorded by the SBC software at the end of the slab compaction as demonstrated in Table 3. At the time of writing this paper, these authors did not find any literature publications for PA mixes in relation to this phenomenon, but similar observations were reported on DG mixes. The difference observed in DG mixes was not as pronounced as in PA. Priyadharshini and Krishnan (2014) compacted different types of DG mixes using SBC and then extracted three 100 mm in diameter and 150 mm in height specimens per slab. The slabs were compacted to different target AV: 4.5, 6.5 and 8.5%. The AV measured in the extracted specimens were 2.5% lower than the target AV: 2, 4, and 6%, respectively. Qiu et al. (2012) reported that the average AV in the extracted specimens from SBC slabs varied from 3.21 to 4.79% depending on the mix while the target AV for slab compaction was 5%. So, it is up to 1.8% lower in the measured AV in the extracted specimens compared to the target AV used for slab compaction.

There are several possible explanations why AV content in the extracted specimens are lower compared to the target AV (as well as AV content recorded by the SBC software at the end of slab compaction). Firstly, during the slab compaction process, the asphalt mix is not subjected to shear stress in the areas within 20 mm from the mold edge along the length of the mold (Qiu et al. 2012). These areas of the slab are likely to have higher AVs compared to the rest of the slab. During the DM specimen extraction from PA slabs, these areas were avoided. Secondly, the SBC compactor was not designed to pre-heat the mold prior to mix loading. PA mix has significantly higher AV content compared to DG mixes. When the PA mix is in contact with the cold mold during mix loading, the rate of mix cooling in the contact areas would be much higher compared to DG mixes, and as a result the edge effect can be more pronounced in PA mixes compared to DG mixes. Thirdly, the cored surface of PA specimens is smoother compared to the untrimmed surface of the slab, which will likely result in lower AV results. And finally, the SBC was originally designed for compacting DG mixes and not necessarily PA mixes.

The negative aspect of using the SBC equipment is that it requires significantly more mix quantities and materials (aggregates, binder and etc.) during the compaction trials to achieve the desired AV in the specimens compared to gyratory compacted specimens. Also, the SBC machine cannot be used for preparation of other slabs while the compacted slab is cooling in the mold. This mold is not detachable from the machine as in the case of gyratory molds. On the positive side, the SBC compaction process is significantly faster (1–4 min/slab) compared to several individually gyratory compacted specimens. There are no special resource requirements needed for compaction of asphalt mixes using the SBC compared to gyratory compaction. The installation and operation requirements such as power and compressed air (dry, clean air at minimum of 600 kPa) are similar. It is a one-person operation with less manual handling involved: one SBC slab versus four individually gyratory compacted specimens for DM testing, for example. A trolley is used to load the mix material into the SBC mold and for slab handling.

5 Conclusions

In this study, 64 specimens of 10 mm NMAS PA mixes with different aggregates and bituminous binders were extracted from SBC slabs. Specimens were cored and trimmed to a target of 100 mm in diameter and 150 mm in height for AV measurements. The following conclusions can be drawn:

- The developed PA sample preparation allows extraction of the specimens with minimum damage. This is critical for performance testing such as DM as well as for AV consistency and homogeneity in the specimens.
- Incorporation of vacuum drying is essential for PA specimen drying. Fan drying alone was not sufficient. PA mixes made with highly vesicular aggregates such as melter slag required more drying time.
- Variability in the replicates within one slab was within ±0.5% from the average AV value. The total variability between all specimens tested was within 2%.

- The slab height was variable, ranging from 163 to 230 mm, but it didn't affect the consistency of the AV in the replicates. The slab height of 183–190 mm was found to be the best for 10 mm NMAS PA mixes in relation to specimen handling. It is important take into the account the NMAS of PA mix when determining the optimum slab height for trimming to the target specimen height and AV content.
- The SBC is capable of producing large slabs of PA mixes for extractions of test replicates with very limited variation in the AV values. This means DM testing can be performed reliably on the PA specimens using the same limits in AV variation as in DG mixes, such as ±0.5% from the average AV content.
- The extracted specimens from SBC are true replicates as they are compacted under the same conditions compared to gyratory compaction (or Marshall) where samples are compacted individually.
- The extracted PA specimens had 3–4% lower AV compared to the slab AV recorded at the end of compaction by the machine. This means SBC trials need to be performed first to determine the target AV content (or density) needed for slab compaction to achieve the desired AV content in the extracted specimens.
- The importance of having pre-heated mold should not be underestimated. PA mixes have higher AV content compared to DG mixes, so the cooling rate of the mix is highly likely will be faster when in contact with the cold mold. It is not a good practice to overheat PA mixes to compensate for a cold mold. It is recommended to include the pre-heating of the SBC mold to the compaction temperature prior to introduction of the mix into it as a compulsory step in the ASTM D7981.
- The height of 185 mm is not a limitation for compaction of PA slabs using SBC. The maximum slab height possible depends on the mix target AV, maximum theoretical density, and equipment limitations. Use of the sample chute should be optional to allow slab compaction to the possible maximum height when needed to maximize the slab use. Good results in relation to AV variability within one slab was achieved without using the loading chute. Slab preparation and handling are different for PA compared to DG mixes. Modifications and improvements to the ASTM D7981 are recommended to take these points into the account for PA mixes.

It is recommended that a comparative study between SBC specimens and field compacted specimens of PA is carried out using, preferably, a non-destructive technique such as CT scanning. This will allow an understanding of the differences in aggregate orientation in these two compaction scenarios. The compaction technique in the field is different for PA mixes compared to DG mixes. The minimum compaction effort is required to achieve the stone-on-stone interlock for PA compared to DG mixes. Thus, the simulation is of static rolling only with no vibratory or oscillatory components.

Lastly, testing of the AV distribution within the specimens extracted from the SBC needs to be investigated to understand the AV variability, or alternatively a CT scanning can be performed on SBC slabs to avoid complicating effects of coring and sawing.

References

AASHTO: AASHTO PP 60-09: Standard Practice for Preparation of Cylindrical Performance Test Specimens Using the Superpave Gyratory Compactor (SGC). AASHTO, Washington, DC (2009)

AASHTO: AASHTO TP 79-13: Standard Method of Test for Determining the Dynamic Modulus and Flow Number for Asphalt Mixtures Using the Asphalt Mixture Performance Tester (AMPT). American Association of State and Highway Transportation Officials, Washington, DC (2013)

ASTM International: ASTM D3203/D3203M-11: Standard Test Method for Percent Air Voids in Compacted Dense and Open Bituminous Paving Mixtures. ASTM International, West Conshohocken, PA (2011a). https://doi.org/10.1520/D3203_D3203M-11

ASTM International: ASTM D7227/D7227M-11: Standard Practice for Rapid Drying of Compacted Asphalt Specimens Using Vacuum Drying Apparatus. ASTM International, West Conshohocken, PA (2011b). https://doi.org/10.1520/D7227_D7227M-11

ASTM International: ASTM D7981-15: Standard Practice for Compaction of Prismatic Asphalt Specimens by Means of the Shear Box Compactor. ASTM International, West Conshohocken, PA (2015). https://doi.org/10.1520/D7981-15

Cooley, L.A., Brumfield, J.W., Mallick, R.B., Mogawer, W.S., Partl, M., Poulikakos, L., Hicks, G.: Construction and maintenance practices for permeable friction courses. NCHRP Report 640, TRB (2009)

Gabrawy, T.: Towards laboratory replication of field compaction: the asphalt shear box compactor. In: Proceedings of the 1st International Conference World of Asphalt Pavements, Sydney, Australia, 20–24 Feb 2000

Goh, S.W., You, Z.: Mechanical properties of porous asphalt pavement materials with warm mix asphalt and RAP. J. Transp. Eng. **138**(1), 90–97 (2012). https://doi.org/10.1061/(ASCE)TE.1943-5436.0000307

Hamzah, M.O., Hasan, M.R.M., Van De Ven, M., Yahaya, A.S.: The effects of initial conditioning and ambient temperatures on abrasion loss and temperature change of porous asphalt. Constr. Build. Mater. **29**, 108–113 (2012). https://doi.org/10.1016/j.conbuildmat.2011.08.050

Holleran, G., Holleran, I., Bearsley, S., Dubois, C.J., Wilson, D.: Epoxy asphalt for durability of open graded mixes: part 1 performance approaches. In: 17th AAPA International Flexible Pavements Conference, Australian Asphalt Paving Association, Melbourne, 13–16 Aug 2017

Kandhal, P.S.: Design, Construction, and Maintenance of Open-Graded Asphalt Friction Courses. Asphalt Institute, National Asphalt Pavement Association (2002)

Molenaar, A.A.A., van de Ven, M.F.C., Rickards, I.: Laboratory compaction of asphalt samples using the shear box compactor. In: 13th AAPA International Flexible Pavements Conference, Australian Asphalt Paving Association, Queensland, Australia, 11–14 Oct 2009

NZ Transport Agency: TNZ P/23:2005 Performance Based Specification for Hotmix Asphalt Wearing Course Surfacing. NZ Transport Agency, New Zealand (2005)

NZ Transport Agency: TNZ P/11:2007 Specification for Open Graded Porous Asphalt. NZ Transport Agency, New Zealand (2007)

NZ Transport Agency: NZTA M1-A: 2016 Performance Specification for Asphalt Binders. NZ Transport Agency, New Zealand (2016)

Priyadharshini, R.Y., Krishnan, J.M.: How workable are modified binder mixtures? A comparative study with gyratory and shear compactor. In: Colloquium on Transportation Systems Engineering and Management, CTR, CED, NIT Calicut, India, 12–13 May 2014

Qiu, J., Li, N., Pramesti, F.P., Van De Ven, M.F.C., Molenaar, A.A.A.: Evaluating laboratory compaction of asphalt mixtures using the shear box compactor. J. Test. Eval. **40**(5) (2012). https://doi.org/10.1520/jte20120070

Standards Australia: AS 2891.2.2-1995. Methods of Sampling and Testing Asphalt. Method 2.2: Sample Preparation—Compaction of Asphalt Test Specimens Using a Gyratory Compactor. Australia (1995)

Zhang, J., Alvarez, A.E., Lee, S.I., Torres, A., Walubita, L.F.: Comparison of flow number, dynamic modulus, and repeated load tests for evaluation of HMA permanent deformation. Constr. Build. Mater. **44**, 391–398 (2013). https://doi.org/10.1016/j.conbuildmat.2013.03.013

Concrete Surface Hardener Laboratory Performance Study for Caltrans

DingXin Cheng(✉), Madelyn L. Giles, and Jarvis Thor

California Pavement Preservation Center, Chico, CA 95929-0603, USA
{dxcheng,mlgiles}@csuchico.edu,
jthor3@mail.csuchico.edu

Abstract. California Department of Transportation (Caltrans) is interested in preserving the concrete pavement surface from the wear and abrasion of traffic loading. Especially, on a portion of concrete pavement on Interstate Highway 80 near Truckee area where the pavement deteriorates at an accelerated rate due to snow chains and heavy traffic loading in the truck lanes. Historically, Caltrans has tried thin hot mix asphalt overlay, bonded wearing course, rubberized hot mix asphalt overlay, and micro-surfacing as a fix for these wear problems with various levels of success. Currently, Caltrans is looking for new cost effective concrete surface preservation treatments to solve the wear issues. Concrete surface hardeners may offer a less expensive preservation alternative compared to the previous solutions. There were five different concrete surface hardeners from three companies selected for the laboratory study including Transil Plus and Pentra Sil (IH) from Convergent Concrete, Nano-Lithium 25 from Advan, CS and CLS densifiers from Trinic. Three different tests were performed on the treated and untreated cement mortar surfaces. The concrete surface abrasion test, Caltrans Test Method 550, was used to evaluate the surface abrasion resistance of specimens; Mohs hardness test was used to study the surface hardness changes during the surface hardener treatments; and the rebound hammer test, ASTM C805, was used to evaluate the strengths of various treated and untreated specimens. The statistical testing results were calculated and summarized to show the effectiveness of the surface hardeners on various cement types with different water to cement ratios. The test results showed the levels of effectiveness of different concrete surface hardeners had on the various mix designs.

Keywords: Concrete pavement preservation · Surface hardener
Concrete abrasion

1 Introduction

Caltrans is interested in preserving concrete pavement from the wear and abrasion of traffic loading, snow chains, and studded tires. Historically Caltrans has tried thin overlay, thin bonded wearing course and micro surfacing as a fix for these problems with mixed levels of success. Currently, Caltrans is looking for new cost effective surface preservation treatments for preventative maintenance to solve the deterioration issues. Lithium-based concrete surface hardeners have long been utilized to protect concrete floors (Nasvik 2008). There were field test sections both with and without

Z. Hossain et al. (eds.), *Solving Pavement and Construction Materials Problems with Innovative and Cutting-edge Technologies*, Sustainable Civil Infrastructures,
https://doi.org/10.1007/978-3-319-95792-0_9

concrete surface hardeners in the past (Stokes 2010; Haworth 2011). Concrete surface hardeners are less expensive compared to the previous concrete preventative maintenance alternatives (Riemer et al. 2012). Therefore Caltrans wants both field trial and laboratory testing to validate the effectiveness of various surface hardeners. Of the concrete surface hardeners available today, the laboratory study performed by the CP2 Center tested the following five surface hardener types:

- Convergent Transil Plus
- Convergent Pentra Sil (IH)
- Advan Nano-Lithium 25
- Trinic CS Densifier
- Trinic CLS Densifier.

1.1 Objectives

The purpose of the study is to evaluate the abrasion resistance of various concrete surface hardener independent of coarse aggregate type through specimen preparation, laboratory testing, data analysis, reporting, and coordination with Caltrans, industry, and academia. The primary objectives are:

1. Prepare multiple specimens for each type of mortar and concrete surface hardener using California Test (CT) 551 Part 3 (Caltrans 2012).
2. Determine the average specimen hardness using Mohs hardness test, and the rebound test, ASTM C805 (ASTM 2013).
3. Determine the abrasion resistance for treated and untreated cement mortar surfaces using California Test 550 (Caltrans 2013).
4. Determine the effectiveness of various lithium silicate and colloidal silica concrete surface hardener products for increasing abrasion resistance.

2 Laboratory Testing

These include mix design trial batching, large concrete batching, cylinder fabrication, curing, cutting of specimens, surface hardener application and required laboratory abrasion testing.

2.1 Materials

Several cementitious materials were used in the study, including Lehigh White Type II/V Cement, Sulfate Resistant, TXI Type III Cement, CTS Rapid Set CSA Cement, slag, and fly ash. The aggregates used during batching consisted of a standard ¼" minus concrete sand.

Several different additives were used in the mix designs for this study, which include the following: Darex II Air Entraining Admixture (AEA), MasterAir AE 200 (Air Entraining Admixture), MasterSet Delvo, and MasterGlenium 7500.

The surface hardeners used in the study were the following:

- Convergent Transil Plus (9% modified lithium)—A clear, penetrating, breathable, surface-applied liquid treatment which strengthens, protects and hardens concrete pavements in cold weather climates subjected to freeze-thaw cycling, studded tires, chains, snow plows, and deicers. This product has a low viscosity surface treatment that penetrates the wear layer of concrete, where it reacts with the free lime to form C–S–H (calcium silicate hydrate) an insoluble wear and moisture protective surface. It combines patented lithium silicate chemistry with reactive silicon catalysts.
- Convergent Pentra Sil (IH) (23% Modified Lithium)—This impregnating hardener is a penetrating Nano-Lithium treatment for concrete surfaces. This environmental friendly treatment also reduces alkalinity (efflorescence) and improves overall abrasion resistance. Its deep-penetrating, non-soluble characteristics provide for a longer life and reduce maintenance cost of concrete surfaces. Pentra Sil (IH) will not contribute to alkali-silica reactivity (ASR) and it inhibits alkali salts (efflorescence) and sweating in concrete surfaces.
- Advan Nano-Lithium 25 (23% Bulk Lithium)—This product has a higher concentration of lithium and is preferred to sodium and potassium and/or lithium silicates for many applications, including concrete surface treatments, specialty paints, and coatings. The reactive silicate allows for improved interaction and compounding when used as a hardener and/or reactive binder in coatings, treatments, and sealers. The unique composition helps formation of a film that increases water resistance, abrasion resistance, hardness, and bonding.
- Trinic CS Densifier (Colloidal Silica)—This product is a Colloidal Silica which is pure silica with a particle 100 times finer than silicates and 1000 times finer than silica fume (which is finer than tobacco smoke), silica fume is 100 times finer than Portland cement which is the rough equivalent size of the grit on 300 grit sand paper. This is why it reacts quickly when used as a surface densifier (it begins to gel almost upon contact).
- Trinic CLS Densifier (Colloidal Silica-Lithium Blend)—This product is a blend of Trinic LS Lockup (a lithium silicate based product) and Trinic CS Densifier (a colloidal silica based product).

2.2 Equipment

There is a wide variety of equipment used for the concrete abrasion study. From trial batching to big batching concrete mix designs, creating the specimen to be tested and the actual testing of the specimen, many different pieces of equipment were used for this study. The following is a list of equipment and tools that were utilized to perform this study: concrete mixer, various concrete mixing tools, lime curing bath, flow table and frame, flow table molds, measure mold air void, abrasion testing mold, CT 550 Concrete Abrasion Testing Machine, Mohs hardness test equipment, and ASTM C805 rebound hammer.

2.3 Mix Design

The design criteria for the cement mortar mix required 10–12% air content and a flow of 75–85 mm. Several trial batches for each mix design were produced before producing a larger batch of 4 in. diameter by 8 in. height cylinders. Some of the main components in the mix design are sand to cement ratio, water to cement ratio, amount of air entraining agent (AEA), water reducing admixture (Glenium 7500), and set retardant (Delvo).

For this study, ASTM C305, Mechanical Mixing of Hydraulic Cement Pastes and Mortars of Plastic Consistency procedure was used. The mix designs were determined by starting with industry recommended material/admix proportioning and then finalized by trial batching to meet the air voids and flow requirements.

Table 1 shows the final mix design proportions developed for each of the mortar type that met the required air voids and flow that will be used to test for the various required concrete surface hardeners.

3 Laboratory Test Results and Analysis

Prior to the abrasion testing, the prepared mortar specimens were treated with the specified concrete surface hardeners. The amount or application rate of concrete surface hardener applied to the specimen for abrasion testing has an application range provided by each manufacturer. They range from 150 to 600 ft^2/gallon. The manufacturer's recommendation for applying the surface hardeners also states to keep the surface wet 15–20 min, but without letting it puddle up on the surface. Table 2 shows the details of surface hardener application rates on difference cement mortars including varied water cement ratios.

3.1 Abrasion Testing Results

The abrasion test results are organized by type of cement used in the mix design and the water to cement ratio. The results from both the three one-minute abrasion testing performed and also results from the straight three minute abrasion testing are saved. For the three one-minute tests, after the abrasion test finished one minute, take the specimen out and measure the concrete surface mass loss. And then repeat the above process for two more times. For the three minutes abrasion testing, the abrasion test will not stop until continuous three minute abrasion testing is completed.

As an example, the Type II cement mortar specimens are used here to illustrate the organization of the results. Test results for Type II Cement with a water to cement ratio w/c of 0.4 and 0.44 are shown in Tables 3 and 4. The results show that the surface hardener, Transil Plus, lost the least material through abrasion in each of the 1 min tests and also cumulatively. Transil Plus performed best for all concrete surface hardener products tested and the corresponding control.

To further study the effectiveness of the surface hardener on Type II cement, six separate 30 s abrasion tests were conducted. One set was our control set without surface hardener, and the other set of three specimen had the Transil Plus surface hardener applied according to manufacturers recommendations. We found a significant difference in the material loss in the first 30 s tests, but in each successive 30 s test, the material loss of the Transil Plus specimen became more comparable to the loss of the Control specimen. Table 5 shows an isolated testing sequence that was performed.

Table 1. Mix design for different types of cement mortars

Batch date	Mortar type	Sand cement ratio	w/c	AEA (lb)	Delvo Set Ret. (lb)	Glenium Wtr Red. (lb)	Fly ash (lb)	Slag (lb)	Cement (lb)	Sand (lb)	Water (lb)	Mix time	% air void	Flow
12/31/2014	Type II	1.7	0.4	0.2490	0	0	0	0	40.8266	69.7520	17.0247	10 min	9.5	79.7
1/2/2015	Type II	2.1	0.44	0.1915	0	0	0	0	35.3272	74.6090	16.2350	10 min	8.9	84.5
4/17/2015	Type II Slag	1.7	0.4	0.2336	0	0.1349	0	19.7133	19.7133	67.2820	16.5190	30 min	10.0	107.0
4/17/2015	Type II Slag	1.7	0.44	0.2269	0	0	0	19.1480	19.1480	65.4130	17.5170	1 h	10.0	89.0
2/13/2015	Type II fly ash	2	0.4	0.2163	0	0	8.8669	0	26.6008	71.2690	14.9180	10 min	10.1	79.5
2/23/2015	Type II fly ash	2.3	0.44	0.1349	0	0	8.2936	0	24.8808	76.6630	15.3800	10 min	10.3	88.6
3/20/2015	Type III	1.5	0.4	0.2560	0	0	0	0	41.9729	62.9600	17.3620	9.5 min	12.1	93.0
4/20/2015	Type III fly ash	1.7	0.4	0.0761	0	0.1793	9.3575	0	28.0724	63.9690	15.5890	10 min	10.2	92.0
4/10/2015	CSA	1.5	0.4	0.2563	0.4387	0.5751	0	0	42.0221	63.3360	17.4520	15 min	13.2	95.5

Table 2. Surface hardener application rates

Modifier type	Densifter names	Mix ratio (water/product)	Manufacture's recommended application range	App rate (ft^2/gal)	Mass applied (g)	Type II 0.4	Type II 0.44	Type II 0.4 w/slag	Type II 0.44 w/slag	Type II 0.4 w/fly ash	Type II 0.44 w/fly ash	Type III 0.4	Type III w/fly ash 0.4	CSA 0.4
9% Modified Li	Convergent Transil Plus	0	150–200 ft^2/gal	200	1.74	✓	✓	–	–	–	–	–	–	–
				150	2.33	–	–	✓	✓	✓	✓	✓	✓	✓
23% Modified Li	Convergent Pentra Sil (IH)	0	400–500 ft^2/gal	400	0.85	–	–	✓	✓	✓	✓	✓	✓	✓
				213	1.60	✓	✓	–	–	–	–	–	–	–
23% Bulk Li	Advan Nano-Lithium 25	0	Start with 450, If dry within 18 min, add more to keep wet until	450	0.86	–	–	✓	✓	✓	✓	✓	✓	✓
				227	1.70	✓	✓	–	–	–	–	–	–	–
Colloidal Silica	Trinic CS densifier	4	400–600 ft^2/gal	400	0.89	–	–	✓	✓	✓	✓	✓	✓	✓
				328	1.40	✓	✓	–	–	–	–	–	–	–
Colloidal Silica-Li Blend	Trinic CLS densifier	4	100–600 ft^2/gal	400	0.83	–	–	✓	✓	✓	✓	✓	✓	✓
				248	1.41	✓	✓	–	–	–	–	–	–	–

Table 3. Type II w/c 0.4 surface hardener test results

Cement type: Type II cement																								
	Hardener treatment																							
	Transil Plus				Pentra Sil (IH)				Nano Lithium 25				Trinic CS				Trinic CLS				None			
	9% Li				23% Li				25% Bulk Li				Colloidal Silica				CLS Blend				Control			
w/c ratio	0.4				0.4				0.4				0.4				0.4				0.4			
Three 1-min test	**s1**	**s2**	**s3**	**Avg**	**s1**	**s2**	**s3**	**Avg**	**s1**	**s2**	**s3**	**Avg**	**s1**	**s2**	**s3**	**Avg**	**s1**	**s2**	**s3**	**Avg**	**s1**	**s2**	**s3**	**Avg**
starting Wt	833.6	837.8	847.4	839.6	832.9	867.6	795.9	832.1	856.0	818.5	844.4	839.6	864.7	834.6	816.6	838.6	841.8	869.7	860.2	857.2	852.4	814.3	850.1	838.9
cum 1 min lost	3.7	3.7	3.7	3.7	4.6	4.1	4.6	4.4	4.5	4.1	4.5	4.4	4.9	5.3	4.6	4.9	5.0	5.2	5.1	5.1	4.4	5.4	4.8	4.9
cum 2 min lost	9.2	9.2	9.5	9.3	10.0	9.6	10.2	9.9	10.1	9.4	10.1	9.9	11.1	11.1	10.0	10.7	9.7	10.6	11.0	10.4	9.9	10.3	10.3	10.2
cum 3 min lost	13.2	13.1	13.7	13.3	14.5	14.0	13.9	14.1	14.7	13.8	14.6	14.4	16.1	15.7	14.2	15.3	14.1	14.9	15.4	14.8	14.4	14.5	14.2	14.4
3 min test	**s1**	**s2**	**s3**	**Avg**	**s1**	**s2**	**s3**	**Avg**	**s1**	**s2**	**s3**	**Avg**	**s1**	**s2**	**s3**	**Avg**	**s1**	**s2**	**s3**	**Avg**	**s1**	**s2**	**s3**	**Avg**
starting Wt	851.2	831.6	832.1	838.3	819.9	836.0	837.1	831.0	852.5	847.1	853.2	850.9	846.0	843.7	840.4	843.4	846.0	846.0	836.6	842.9	833.1	830.8	833.2	832.4
3 min test total lost	13.2	13.1	13.5	13.3	13.1	13.7	14.2	13.7	16.6	14.5	14.2	15.1	15.2	12.8	13.0	13.7	15.2	12.9	10.7	12.9	13.9	13.8	13.8	13.8

Note Results obtained using California Test (CT) 550

Table 4. Type II w/c 0.44 surface hardener test results

Cement type: Type II cement															
	Hardener treatment														
	Transil Plus							Pentra Sil (IH)				Nano Lithium 25			
	9% Li							23% Li				25% Bulk Li			
w/c ratio	0.44							0.44				0.44			
Three 1-min test	**s1**	**s2**	**s3**	**s4**	**s5**	**s6**	**Avg**	**s1**	**s2**	**s3**	**Avg**	**s1**	**s2**	**s3**	**Avg**
starting Wt	823.8	837.1	859.4	800.6	830.4	813.0	827.4	844.2	819.5	845.3	836.3	818.4	839.0	853.3	836.9
cum 1 min lost	3.4	5.3	4.2	2.6	4.5	4.0	4.0	5.2	7.6	4.2	5.7	4.3	4.7	4.5	4.5
cum 2 min lost	9.3	10.6	10.1	7.3	9.5	8.5	9.2	11.0	13.2	10.3	11.5	9.9	10.9	10.3	10.4
cum 3 min lost	13.1	14.7	14.4	10.5	13.2	12.3	13.0	15.2	17.7	14.7	15.9	14.3	14.9	14.6	14.6
3 min test	**s1**	**s2**	**s3**	**–**	**–**	**–**	**Avg**	**s1**	**s2**	**s3**	**Avg**	**s1**	**s2**	**s3**	**Avg**
starting Wt	812.5	845.2	817.3	–	–	–	825.0	823.1	853.4	854.5	843.7	832.3	865.8	780.9	826.3
3 min test total lost	13.7	14.7	11.0	–	–	–	13.1	14.1	15.0	11.7	13.6	13.6	14.8	10.9	13.1
	Trinic CS				Trinic CLS				None						
	Colloidal Silica				CLS Blend				Control						
w/c ratio	0.44				0.44				0.44						
Three 1-min test	**s1**	**s2**	**s3**	**Avg**	**s1**	**s2**	s3	Avg	s1	s2	s3	s4	s5	s6	Avg
starting Wt	831.2	834.6	847.8	837.9	847.6	853.5	846.4	849.2	831.2	840.6	826.9	821.0	805.1	830.9	826.0
cum 1 min lost	4.7	4.7	4.5	4.6	4.7	1.7	5.8	4.1	4.7	4.6	4.9	4.5	4.5	5.6	4.8
cum 2 min lost	10.3	10.2	10.1	10.2	10.1	8.0	10.3	9.5	10.3	10.2	10.5	10.8	9.2	11.3	10.4
cum 3 min lost	14.3	14.6	13.5	14.1	14.2	13.0	14.0	13.7	14.3	14.0	14.8	15.1	13.3	15.1	14.4
3 min test	**s1**	**s2**	**s3**	**Avg**	**s1**	**s2**	**s3**	**Avg**	**s1**	**s2**	**s3**	**–**	**–**	**–**	**Avg**
starting Wt	852.9	837.1	833.0	841.0	834.4	840.9	848.6	841.3	842.4	816.6	844.4	–	–	–	834.5
3 min test total lost	14.0	14.0	11.7	13.2	12.9	14.6	13.8	13.8	14.0	13.5	13.6	–	–	–	13.7

Note Results obtained using California Test (CT) 550

Table 5. Type II w/c 0.44 30 s interval test results

Cement type: Type II cement								
	Hardener treatments							
	Transil Plus				None			
	9% Li				Control			
w/c ratio	0.44				0.44			
6–30 s test	s1	s2	s3	Avg	s1	s2	s3	Avg
starting Wt	833.8	859.3	846.4	346.5	843.2	827.8	836.5	835.8
cum 30 s lost	0.1	0.1	0.2	0.1	2.1	1.2	1.0	1.4
cum 30 s lost	2.9	3.0	3.3	3.1	6.8	4.0	3.6	4.8
cum 30 s lost	6.3	6.9	6.3	6.5	9.4	6.8	6.3	7.5
cum 30 s lost	8.9	9.8	3.5	9.1	11.2	8.9	8.2	9.4
cum 30 s lost	11.0	12.6	10.4	11.3	13.3	10.8	9.8	11.3
cum 30 s lost	13.0	15.1	12.1	13.4	15.5	12.4	11.3	13.1

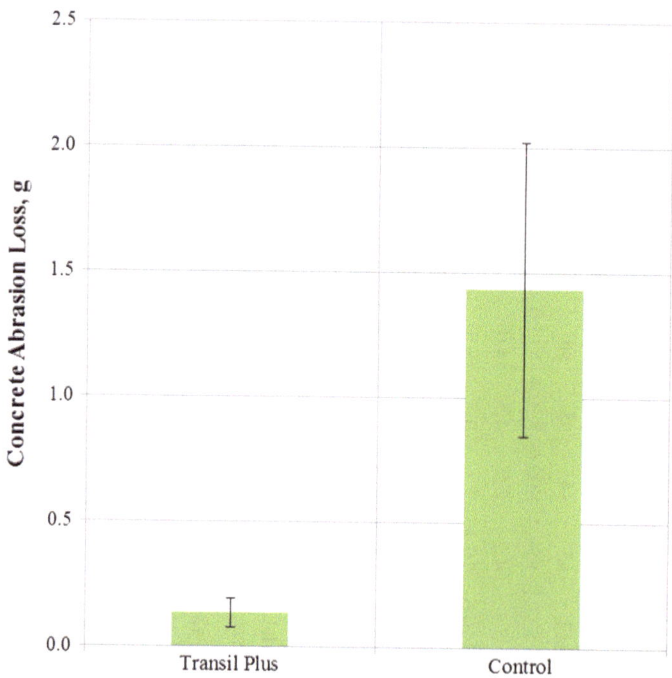

Fig. 1. CT 550 abrasion testing results for Type II Cement with 0.44 w/c ratio 30 s intervals

Figure 1 shows the comparison of Transil Plus and control specimens with the amount of mass that was lost in the first 30 s test for Type II cement with 0.44 w/c ratio. This graph depicts the difference between Transil Plus and the Control specimen.

Clearly the Transil Plus concrete surface hardener made a significant difference in the first 30 s of testing compared to the Control specimen that had no surface hardener applied.

3.2 T-Test Comparison on Abrasion Testing Results

To evaluate the effectiveness of the concrete surface hardener treatments, t-tests were conducted to compare various surface hardener treatments with control cement mortar surfaces. The statistical analysis was conducted at 95% confident level with Type I error alpha was equal to 0.5. The number of replicates for each treatment is 3. If a calculated t-statistics is greater than the one-tail critical t value, the Null hypothesis can be rejected and the concrete surface hardener improves the performance significantly.

As an example, Type II cement is selected as an illustration. Table 6 shows the results from the t-tests performed on the Type II w/Slag with 0.4 w/c ratio abrasion tests results. None of the specimens in the tests showed significant improvement when compared to the Control specimens.

Table 6. T-test results for Type II with Slag with 0.4 w/c ratio abrasion test

Product	Cement type	w/c ratio	Time total 1 min/3 min	t-Statistic	t-Critical one-tail	Significant or not on Improvement Yes/No
Transil Plus	Type II w/Slag	0.4	1	2.744	2.920	No
Transil Plus	Type II w/Slag	0.4	3	0.667	2.353	No
Pentra Sil	Type II w/Slag	0.4	1	−0.442	2.920	No
Pentra Sil	Type II w/Slag	0.4	3	0.667	2.353	No
Nano Lith	Type II w/Slag	0.4	1	0.555	2.353	No
Nano Lith	Type II w/Slag	0.4	3	0.000	2.353	No
Trinic CS	Type II w/Slag	0.4	1	0.200	2.920	No
Trinic CS	Type II w/Slag	0.4	3	0.621	2.353	No
Trinic CLS	Type II w/Slag	0.4	1	1.265	2.920	No
Trinic CLS	Type II w/Slag	0.4	3	0.307	2.132	No

Table 7 shows the results from the t-tests performed on the Type II w/Slag with 0.44 w/c ratio abrasion tests results. The Transil Plus specimens in the three 1-min tests and also the 3-min tests showed significant improvement when compared to the Control specimens.

Table 7. T-test results for Type II with Slag with 0.44 w/c ratio abrasion test

Product	Cement type	w/c ratio	Time total 1 min/3 min	t-Statistic	t-Critical one-tail	Significant or not on Improvement Yes/No
Transil Plus	Type II w/Slag	0.44	1	3.443	2.920	Yes
Transil Plus	Type II w/Slag	0.44	3	2.859	2.132	Yes
Pentra Sil	Type II w/Slag	0.44	1	1.594	2.132	No
Pentra Sil	Type II w/Slag	0.44	3	−0.136	2.132	No
Nano Lith	Type II w/Slag	0.44	1	1.391	2.920	No
Nano Lith	Type II w/Slag	0.44	3	0.407	2.132	No
Trinic CS	Type II w/Slag	0.44	1	1.148	2.920	No
Trinic CS	Type II w/Slag	0.44	3	0.646	2.353	No
Trinic CLS	Type II w/Slag	0.44	1	0.804	2.132	No
Trinic CLS	Type II w/Slag	0.44	3	0.632	2.920	No

3.3 Rebound Hammer Testing Results

The following section shows the rebound hammer testing results inserted into Table 8 for the rebound hammer test on: Type II 0.44, Type II 0.4, Type II Flay Ash 0.4, Type III 0.4, CSA 0.4, Type II w/Slag 0.4, Type II w/Slag 0.44, Type III w/Flay Ash 0.4, and Type II w/Fly Ash. There also graphs included for each concrete batch type showing the rebound hardness number for each specific concrete surface hardener applied.

3.4 T-Test Comparison on Rebound Hammer Testing Results

To evaluate the effectiveness of the surface hardener treatments, t-tests were conducted to compare various surface hardener treated with control cement mortar surfaces. The statistical analysis was conducted at 95% confident level with Type I error alpha was equal to 0.5. The number of replicates for each treatment is 3. If a calculated t-statistics

Table 8. ASTM C805 rebound hammer testing results

Material	Surface hardener	Rebound #	Test 1	Test 2	Test 3	Test 4	Test 5	Test 6	Test 7	Test 8	Test 9	Test 10	Stad. Dev.
Type II 0.4	Transil Plus	39.6	40	43	40	43	36	40	40	41	33	40	3.0
Type II 0.4	Pentra Sil	43.2	39	42	44	47	40	43	43	45	44	45	2.4
Type II 0.4	Nano Lithium 25	42.0	42	45	41	42	41	41	44	41	43	40	1.6
Type II 0.4	Trinic CS	35.3	37	34	34	39	35	33	35	33	32	41	2.9
Type II 0.4	Trinic CLS	36.3	39	40	34	38	30	35	36	35	36	40	3.1
Type II 0.4	Control	33.8	30	36	29	29	35	40	34	40	32	33	4.0
Material	**Surface hardener**	**Rebound #**	**Test 1**	**Test 2**	**Test 3**	**Test 4**	**Test 5**	**Test 6**	**Test 7**	**Test 8**	**Test 9**	**Test 10**	**Stad. Dev.**
Type II 0.44	Transil Plus	31.8	31	31	33	32	34	33	30	31	32	31	1.2
Type II 0.44	Pentra Sil	31.0	30	31	30	32	32	34	30	31	29	31	1.4
Type II 0.44	Nano Lithium 25	33.8	32	38	38	32	31	34	31	36	34	32	2.7
Type II 0.44	Trinic CS	31.1	32	32	30	32	33	32	29	29	30	32	1.4
Type II 0.44	Trinic CLS	35.9	38	36	34	33	36	34	38	37	35	38	1.9
Type II 0.44	Control	31.8	30	31	30	32	34	28	35	35	30	33	2.4
Material	**Surface hardener**	**Rebound #**	**Test 1**	**Test 2**	**Test 3**	**Test 4**	**Test 5**	**Test 6**	**Test 7**	**Test 8**	**Test 9**	**Test 10**	**Stad. Dev.**
Type II w/slag 0.4	Transil Plus	35.3	30	32	37	37	36	38	37	30	38	38	3.3
Type II w/slag 0.4	Pentra Sil	37.2	42	38	34	37	39	40	37	36	39	30	3.4
Type II w/slag 0.4	Nano Lithium 25	37.5	36	38	36	36	38	41	38	36	40	36	1.8
Type II w/slag 0.4	Trinic CS	36.9	35	36	40	38	37	36	40	37	36	34	2.0
Type II w/slag 0.4	Trinic CLS	35.6	35	35	38	37	33	33	38	35	36	36	1.8
Type II w/slag 0.4	Control	37.5	39	32	38	39	35	38	39	35	39	41	2.7
Material	**Surface hardener**	**Rebound #**	**Test 1**	**Test 2**	**Test 3**	**Test 4**	**Test 5**	**Test 6**	**Test 7**	**Test 8**	**Test 9**	**Test 10**	**Stad. Dev.**

(*continued*)

Table 8. (*continued*)

Material	**Surface hardener**	**Rebound #**	**Test 1**	**Test 2**	**Test 3**	**Test 4**	**Test 5**	**Test 6**	**Test 7**	**Test 8**	**Test 9**	**Test 10**	**Stad. Dev.**
Type II w/slag 0.44	Transil Plus	33.5	36	28	31	30	33	35	38	38	32	34	3.3
Type II w/slag 0.44	Pentra Sil	35.8	33	39	37	36	38	32	36	36		35	2.2
Type II w/slag 0.44	Nano Lithium 25	35.8	30	38	37	36	35	37	40	35	34	36	2.7
Type II w/slag 0.44	Trinic CS	35.3	33	34	36	32	38	39	38	36	34	33	2.5
Type II w/slag 0.44	Trinic CLS	35.8	37	36	39	36	35	36	40	33	31	35	2.6
Type II w/slag 0.44	Control	36.1	38	36	39	38	30	36	32	37	40	35	3.1
Material	**Surface hardener**	**Rebound #**	**Test 1**	**Test 2**	**Test 3**	**Test 4**	**Test 5**	**Test 6**	**Test 7**	**Test 8**	**Test 9**	**Test 10**	**Stad. Dev.**
Type II Fly Ash 0.4	Transil Plus	30.5	32	30	29	28	32	32	30		31	32	1.4
Type II Fly Ash 0.4	Pentra Sil	30.3	29	30	30	30	31	32	30	31	31	30	0.8
Type II Fly Ash 0.4	Nano Lithium 25	30.1	30	31	32	29	30	30	29	30	29	31	1.0
Type II Fly Ash 0.4	Trinic CS	32.9	34	32	33	33	36	32	30	31	35	33	1.8
Type II Fly Ash 0.4	Trinic CLS	29.1	26	32	28	28	29	31	28	31	29	29	1.8
Type II Fly Ash 0.4	Control	30.1	29	31	32	30	29	30	31	29	32	29	1.1
Material	**Surface hardener**	**Rebound #**	**Test 1**	**Test 2**	**Test 3**	**Test 4**	**Test 5**	**Test 6**	**Test 7**	**Test 8**	**Test 9**	**Test 10**	**Stad. Dev.**
Type II Fly Ash 0.44	Transil Plus	29.9	28	30	29	31	30	32	29	32	29	29	1.4
Type II Fly Ash 0.44	Pentra Sil	32.4	31	31	34	33	34	34	31	32	30	33	1.4
Type II Fly Ash 0.44	Nano Lithium 25	33.5	31	36	34	32	34	34	33	33	34	33	1.4
Type II Fly Ash 0.44	Trinic CS	32.9	32	34	33	33	33	34	33	32	34	31	0.8
Type II Fly Ash 0.44	Trinic CLS	32.5	34	31	33	31	33	32	33	33	32	33	0.8
Type II Fly Ash 0.44	Control	29.1	28	30	28	29	31	27	32	29	27	30	1.5
Material	**Surface hardener**	**Rebound #**	**Test 1**	**Test 2**	**Test 3**	**Test 4**	**Test 5**	**Test 6**	**Test 7**	**Test 8**	**Test 9**	**Test 10**	**Stad. Dev.**

(*continued*)

Table 8. *(continued)*

Material	**Surface hardener**	**Rebound #**	**Test 1**	**Test 2**	**Test 3**	**Test 4**	**Test 5**	**Test 6**	**Test 7**	**Test 8**	**Test 9**	**Test 10**	**Stad. Dev.**
Type III 0.4	Transil Plus	34.1	34	36	35	32	33	34	36	30	37		2.2
Type III 0.4	Pentra Sil	39.3	40	38	39	42	39	41	39	36	41	38	1.8
Type III 0.4	Nano Lithium 25	37.5	34	37	37	34	39	39	39	35	39	42	2.6
Type III 0.4	Trinic CS	36.9	37	38	33	38	38	38	36	40	37	34	2.1
Type III 0.4	Trinic CLS	37.8	41	37	38	38	38	39	35	38	37	37	1.5
Type III 0.4	Control	36.1	35	35	34	36	36	41	35	34	40	35	2.4
Material	**Surface hardener**	**Rebound #**	**Test 1**	**Test 2**	**Test 3**	**Test 4**	**Test 5**	**Test 6**	**Test 7**	**Test 8**	**Test 9**	**Test 10**	**Stad. Dev.**
Type III w/Fly Ash 0.4	Transil Plus	32.7	31	30	32	36	33	34	32	32	36	31	2.1
Type III w/Fly Ash 0.4	Pentra Sil	33.9	29	34	35	28	31	34	38	39	38	33	3.8
Type III w/Fly Ash 0.4	Nano Lithium 25	35.2	34	38	34	35	33	37	38	34	32	37	2.1
Type III w/Fly Ash 0.4	Trinic CS	34.7	31	33	35	36	37	35	40	32	33	35	2.6
Type III w/Fly Ash 0.4	Trinic CLS	31.8	30	32	33	32	32	32	31	35	30	31	1.5
Type III w/Fly Ash 0.4	Control	32.05	32	29	36	33	36	30	29.5	35	29	31	2.8
Material	**Surface hardener**	**Rebound #**	**Test 1**	**Test 2**	**Test 3**	**Test 4**	**Test 5**	**Test 6**	**Test 7**	**Test 8**	**Test 9**	**Test 10**	**Stad. Dev.**
CSA 0.4	Transil Plus	32.6	35	26	31	30	29	32	37	37	37	32	3.8
CSA 0.4	Pentra Sil	34.1	35	34	30	34	30	38	32	40	38	30	3.7
CSA 0.4	Nano Lithium 25	33.6	27	33	33	32	36	36	38	35	29	37	3.5
CSA 0.4	Trinic CS	35.4	34	34	32	32	38	40	36	35	38	35	2.6
CSA 0.4	Trinic CLS	33.0	30	29	33	33	35	34	29	34	38	35	2.9
CSA 0.4	Control	33.6	32	32	32	35		30	32	38	34	37	2.7

is greater than the one-tail critical t value, the Null hypothesis can be rejected and the surface hardener improved the performance significantly.

As an example, Table 9 shows a significant improvement for Transil Plus, Pentra Sil and Nano Lith. There was no significant improvement for Trinic CS, or Trinic CLS.

Table 9. T-test results for Type II with 0.4 w/c ratio rebound hammer test

Product	Cement type	w/c ratio	t-Statistic	t-Critical one-tail	Significant or not on Improvement (Yes/No)
Transil Plus	Type II	0.4	3.628	1.740	Yes
Pentra Sil	Type II	0.4	5.916	1.753	Yes
Nano Lith	Type II	0.4	5.973	1.782	Yes
Trinic CS	Type II	0.4	0.956	1.746	No
Trinic CLS	Type II	0.4	1.551	1.740	No

3.5 Mohs Hardness Test Results

The following section shows the raw data inserted into Table 10 for the Mohs Hardness Test on: Type II 0.44, Type II 0.4, Type II Fly Ash 0.4, Type III 0.4, CSA 0.4, Type II w/Slag 0.4, Type II w/Slag 0.44, Type III w/Flay Ash 0.4, and Type II w/Fly Ash.

3.6 T-Test Comparison on Mohs Hardness Test Results

To evaluate the effectiveness of the surface hardener treatments, t-tests were conducted to compare various surface hardener treated with control cement mortar surfaces. The statistical analysis was conducted at 95% confident level with Type I error alpha was equal to 0.5. The number of replicates for each treatment is 3. If a calculated t-statistics is greater than the one-tail critical t value, the Null hypothesis can be rejected and the surface hardener improved the performance significantly. As an example, Table 11 shows a significant improvement for Pentra Sil, and Trinic CS. There was no significant improvement for Transil Plus, Nano Lith, or Trinic CLS.

4 Conclusions and Recommendations

4.1 Conclusions

The following are the conclusions resulting from this concrete surface hardener laboratory performance study:

Table 10. Mohs hardness testing results

Material	Test date	Surface hardener	Mohs test # 1	Mohs test # 2	Mohs test # 3	Avg Mohs #	Stad. Dev.
Type II 0.4	6/15/2015	Transil Plus	6.5	5.5	5.5	5.8	0.6
Type II 0.4	6/15/2015	Pentra Sil	7.0	5.5	6.5	6.3	0.8
Type II 0.4	6/15/2015	Nano Lithium 25	6.5	3.5	7.5	5.8	2.1
Type II 0.4	6/15/2015	Trinic CS	7.5	5.5	6.5	6.5	1.0
Type II 0.4	6/15/2015	Trinic CLS	7.5	4.5	7.5	6.5	1.7
Type II 0.4	6/15/2015	Control	5.5	4.5	4.5	4.8	0.6
Material	**Test date**	**Surface hardener**	**Mohs test # 1**	**Mohs test # 2**	**Mohs test # 3**	**Avg Mohs #**	**Stad. Dev.**
Type II 0.44	5/28/2015	Transil Plus	6.5	5.5	5.5	5.8	0.6
Type II 0.44	5/28/2015	Pentra Sil	7.0	3.5	7.5	6.0	2.2
Type II 0.44	5/28/2015	Nano Lithium 25	6.5	3.5	7.5	5.8	2.1
Type II 0.44	5/28/2015	Trinic CS	5	3.5	5.5	4.7	1.0
Type II 0.44	5/28/2015	Trinic CLS	5.5	4.5	7.5	5.8	1.5
Type II 0.44	5/28/2015	Control	5.5	3.5	5.5	4.8	1.2
Material	**Test date**	**Surface hardener**	**Mohs test # 1**	**Mohs test # 2**	**Mohs test # 3**	**Avg Mohs #**	**Stad. Dev.**
Type II w/slag 0.4	7/28/2015	Transil Plus	5.5	5.5	5.5	5.5	0.0
Type II w/slag 0.4	7/28/2015	Pentra Sil	4.5	4.5	4.5	4.5	0.0
Type II w/slag 0.4	7/28/2015	Nano Lithium 25	4.5	4.5	4.5	4.5	0.0
Type II w/slag 0.4	7/28/2015	Trinic CS	4.5	4.5	4.5	4.5	0.0
Type II w/slag 0.4	7/28/2015	Trinic CLS	4.5	4.5	4.5	4.5	0.0
Type II w/slag 0.4	7/28/2015	Control	5.5	5.5	5.5	5.5	0.0
Material	**Test date**	**Surface hardener**	**Mohs test # 1**	**Mohs test # 2**	**Mohs test # 3**	**Avg Mohs #**	**Stad. Dev.**
Type II w/slag 0.44	7/30/2015	Transil Plus	6	6	6	6.0	0.0
Type II w/slag 0.44	7/30/2015	Pentra Sil	5.0	5.0	5.0	5.0	0.0
Type II w/slag 0.44	7/30/2015	Nano Lithium 25	3.5	3.5	3.5	3.5	0.0
Type II w/slag 0.44	7/30/2015	Trinic CS	3.5	3.5	3.5	3.5	0.0
Type II w/slag 0.44	7/30/2015	Trinic CLS	4.5	4.5	4.5	4.5	0.0

(*continued*)

Table 10. (*continued*)

Material	Test date	Surface hardener	Mohs test # 1	Mohs test # 2	Mohs test # 3	Avg Mohs #	Stad. Dev.
Type II w/slag 0.44	7/30/2015	Control	3.5	3.5	3.5	3.5	0.0
Material	**Test date**	**Surface hardener**	**Mohs test # 1**	**Mohs test # 2**	**Mohs test # 3**	**Avg Mohs #**	**Stad. Dev.**
Type II Fly Ash 0.4	5/20/2015	Transil Plus	5.5	5.5	4.5	5.2	0.6
Type II Fly Ash 0.4	5/20/2015	Pentra Sil	6.0	3.5	7.5	5.7	2.0
Type II Fly Ash 0.4	5/20/2015	Nano Lithium 25	5.5	3.5	7.5	5.5	2.0
Type II Fly Ash 0.4	5/20/2015	Trinic CS	6.5	3.5	5.5	5.2	1.5
Type II Fly Ash 0.4	5/20/2015	Trinic CLS	5.5	4.5	6.5	5.5	1.0
Type II Fly Ash 0.4	5/20/2015	Control	5.5	4.5	5.5	5.2	0.6
Material	**Test date**	**Surface hardener**	**Mohs test # 1**	**Mohs test # 2**	**Mohs test # 3**	**Avg Mohs #**	**Stad. Dev.**
Type II Fly Ash 0.44	6/3/2015	Transil Plus	4.5	3.5	4.2	4.1	0.5
Type II Fly Ash 0.44	6/3/2015	Pentra Sil	5.5	5.8	5.8	5.7	0.2
Type II Fly Ash 0.44	6/3/2015	Nano Lithium 25	4.5	5.5	4.5	4.8	0.6
Type II Fly Ash 0.44	6/3/2015	Trinic CS	5.5	3.5	5.2	4.7	1.1
Type II Fly Ash 0.44	6/3/2015	Trinic CLS	6.5	5.2	6.5	6.1	0.8
Type II Fly Ash 0.44	6/3/2015	Control	5.5	4.8	5.5	5.3	0.4
Material	**Test date**	**Surface hardener**	**Mohs test # 1**	**Mohs test # 2**	**Mohs test # 3**	**Avg Mohs #**	**Stad. Dev.**
Type III 0.4	7/10/2015	Transil Plus	5	5	5	5.0	0.0
Type III 0.4	7/10/2015	Pentra Sil	6.5	7.0	6.5	6.7	0.3
Type III 0.4	7/10/2015	Nano Lithium 25	4	4.5	3.5	4.0	0.5
Type III 0.4	7/10/2015	Trinic CS	6.5	6.5	6.5	6.5	0.0
Type III 0.4	7/10/2015	Trinic CLS	4	4	4	4.0	0.0
Type III 0.4	7/10/2015	Control	5	5.5	6.0	5.5	0.5

(*continued*)

Table 10. (*continued*)

Material	Test date	Surface hardener	Mohs test # 1	Mohs test # 2	Mohs test # 3	Avg Mohs #	Stad. Dev.
Material	**Test date**	**Surface hardener**	**Mohs test # 1**	**Mohs test # 2**	**Mohs test # 3**	**Avg Mohs #**	**Stad. Dev.**
Type III w/Fly Ash 0.4	8/21/2015	Transil Plus	3.5	3.5	3.5	3.5	0.0
Type III w/Fly Ash 0.4	8/21/2015	Pentra Sil	3.5	3.5	2.5	3.2	0.6
Type III w/Fly Ash 0.4	8/21/2015	Nano Lithium 25	2.5	3.5	3.5	3.2	0.6
Type III w/Fly Ash 0.4	8/21/2015	Trinic CS	3.5	3.5	3.5	3.5	0.0
Type III w/Fly Ash 0.4	8/21/2015	Trinic CLS	3.5	3.5	3.5	3.5	0.0
Type III w/Fly Ash 0.4	8/21/2015	Control	2.5	3.5	2.5	2.8	0.6
Material	**Test date**	**Surface hardener**	**Mohs test # 1**	**Mohs test # 2**	**Mohs test # 3**	**Avg Mohs #**	**Stad. Dev.**
CSA 0.4	7/28/2015	Transil Plus	6.5	6.5	6.5	6.5	0.0
CSA 0.4	7/28/2015	Pentra Sil	6.5	6.5	6.5	6.5	0.0
CSA 0.4	7/28/2015	Nano Lithium 25	6.5	6.5	6.5	6.5	0.0
CSA 0.4	7/28/2015	Trinic CS	6.5	6.5	6.5	6.5	0.0
CSA 0.4	7/28/2015	Trinic CLS	6.5	6.5	6.5	6.5	0.0
CSA 0.4	7/28/2015	Control	5.5	5.5	5.5	5.5	0.0

Table 11. T-test results for Type II with 0.40 w/c ratio Mohs hardness test

Product	Cement type	w/c ratio	t-Statistic	t-Critical one-tail	Significant or not on Improvement (Yes/No)
Transil Plus	Type II	0.40	2.121	2.132	No
Pentra Sil	Type II	0.40	2.714	2.132	Yes
Nano Lith	Type II	0.40	0.802	2.920	No
Trinic CS	Type II	0.40	2.500	2.353	Yes
Trinic CLS	Type II	0.40	1.581	2.920	No

- For Type II cement with water to cement ratio of 0.4, the three 1-min abrasion testing results showed that Transil Plus treated specimen lost the least amount of materials in each of the 1 min tests and also cumulatively overall. The 3-min abrasion testing results showed that the product Trinic CLS had the least material loss. However, the Trinic CLS had a larger standard deviation than Transil Plus. Transil Plus performed the best overall. Based on the t-test statistics, only Transil Plus showed significant improvement on surface abrasion resistance for both the three 1 and 3-min tests. Other products did not show the significant abrasion improvement. The rebound hammer testing results showed that all hardeners improved the rebound number. T-test results showed that Transil Plus, Pentra Sil (IH) and Nano Lithium 25 improved the rebound number significantly. The Mohs hardness results showed that all surface hardeners increased the hardness of cement mortar surfaces. T-test results showed that Pentra Sil (IH) and Trinic CS increased the Mohs hardness significantly.
- For Type II cement with water to cement ratio of 0.44, the three 1-min abrasion testing results showed that Transil Plus and Trinic CLS had the least amount of materials loss during the first two minutes, then Transil Plus lost the least amount for the cumulative three minutes. The 3-min abrasion testing results showed that Transil Plus lost the least amount of materials overall. Based on the t-test results, only Transil Plus with three 1-min tests showed significant improvement on surface abrasion resistance. The rebound hammer testing results showed that Nano Lithium 25 and Trinic CLS increased the rebound number while based on the t-test results, both Nano Lithium 25 and Trinic CLS increased the rebound number significantly. The Mohs hardness testing results showed that all hardeners increased the hardness number except Trinic CS. However, based on t-test results, none of the hardeners increased the hardness number significantly.
- For Type II cement with water to cement ratio of 0.44, a 30-s interval for up to 3-min tests were also conducted for Transil Plus and Control. The results showed that there were significant improvement in surface abrasion resistance for the first two 30-s intervals.
- For Type II w/Slag cement with water to cement ratio of 0.4, the three 1-min abrasion test results showed that Transil Plus had the least amount of material loss, while others had similar, larger amounts of material loss. The 3-min abrasion testing results also showed that Transil Plus lost the least amount of materials overall.
- For Type II w/Slag cement with water to cement ratio of 0.44, the three 1-min abrasion testing results showed Transil Plus had the least amount of material loss, followed by Pentra Sil (IH). The 3-min abrasion testing results showed that Transil Plus lost the least amount of materials, followed by Trinic CLS. Based on the t-test results, only Transil Plus showed significant improvement on surface abrasion resistance. The rebound hammer testing results showed that no hardeners increased the rebound number and t-test results verified it. The Mohs hardness testing results showed that Transil Plus and Pentra Sil (IH) increased the hardness number.
- For Type II w/Fly Ash with water to cement ratio of 0.4, the three 1-min abrasion testing results showed Transil Plus had the least amount of material loss, followed by Pentra Sil (IH). The 3-min abrasion testing results showed that none of surface hardeners improved the surface abrasion resistance. Based on the t-test results, only

Pentra Sil (IH) with three 1-min tests showed significant improvement in surface abrasion resistance. The rebound hammer testing results showed that Trinic CLS increased the rebound number, while t-test results show that the improvement was significant. The Mohs hardness testing results showed that Pentra Sil (IH), Nano Lithium 25, and Trinic CLS increased the hardness number. However, the t-test results showed that the improvements were insignificant.

- For Type II w/Fly Ash with water to cement ratio of 0.44, the three 1-min abrasion testing results showed Trinic CLS had the least amount of material loss, followed by Pentra Sil (IH). The 3-min abrasion testing results showed that Transil Plus lost the least amount of materials. However, based on the t-test results, the rebound hammer testing results showed that Pentra Sil (IH), Nano Lithium, Trinic CLS, and Trinic CS increased the rebound number significantly. The Mohs hardness testing results showed that all surface hardeners increased the hardness number. However, t-test results showed that the improvement was insignificant.
- For Type III cement with water to cement ratio of 0.4, the three 1-min abrasion testing results showed Transil Plus had the least amount of materials loss, and all others are similar. The 3-min abrasion testing results showed that Transil Plus lost the least amount of materials. Based on the t-test results, only Transil Plus with three 1-min test showed significant improvement on surface abrasion resistance. The rebound hammer testing results showed that Pentra Sil (IH) and Trinic CLS increased the rebound number while t-test results verified it. The Mohs hardness testing results showed that Pentra Sil (IH) and Trinic CS increased hardness significantly with the t-test results verified it.
- For Type III w/Fly Ash with water to cement ratio of 0.4, the three 1-min abrasion testing results showed Transil Plus had the least amount of materials loss, followed by Trinic CS. The 3-min abrasion testing results showed that Trinic CLS lost the least amount of materials followed by Transil Plus. However, based on t-test results, only Transil Plus with three 1-min tests showed significant improvement in surface abrasion resistance. The rebound hammer testing results showed Nano Lithium 25 and Trinic CS increased the rebound number and the t-test results verified it. The Mohs hardness testing results showed that all hardeners increased the hardness number. However, based on t-test results, none of the hardeners increased the hardness number significantly.
- For CSA cement with water to cement ratio of 0.4, the three 1-min abrasion testing results showed that Pentra Sil (IH) and Transil Plus had the least amount of material loss, followed by Nano Lithium 25. The 3-min abrasion testing results showed that Transil Plus lost the least amount of material but with relatively high standard deviation for these test results. Based on t-test results, Pentra Sil (IH) and Nano Lithium with the three 1-min tests showed significant improvement in surface abrasion resistance. The rebound hammer testing results showed that Nano Lithium and Trinic CS increased the rebound number significantly while the t-test results verified it. The Mohs hardness testing results showed that all surface hardeners increased the hardness number.

4.2 Recommendations

The following are the recommendations from this Caltrans concrete surface hardener abrasion study:

- Because the surface hardeners only work for the concrete surface, the current CT 550 may be too aggressive to evaluate the effectiveness of the concrete surface hardener using mortar. Based on the 30-s interval, 1-min interval, and 3-min tests, the modified 30-s tests showed the largest improvement on surface abrasion resistance. It would be useful to conduct the 30-s tests on all the promising surface hardeners.
- The study used the manufacture recommended surface hardener application rates. It would be very beneficial to test the abrasion by using various surface hardener application rates and by reapplying the concrete surface hardeners between the testing intervals.
- The laboratory study results should be correlated to the Caltrans' field pilot study evaluation. The laboratory study could be used to guide the future field applications rates and application frequencies of concrete surface hardeners.

Acknowledgements. We appreciate the financial support of Caltrans for providing the funding for this important and meaningful study. We would like to extend our gratitude to Doran Glauz of Caltrans who provided continuous technical support to this project. We also appreciate the support from Hector Romero and Robert Hogan from Caltrans Headquarters. We would also like to thank Jeff Koebrick from Convergent Concrete Technologies, Vince Perez and Art Bigelow from CTS Cement, WR Grace, and A & A Concrete Suppliers for their contributions and assistance during the project. We greatly appreciate the student assistants of the CP2 Center for their participation in laboratory testing.

References

ASTM: ASTM C805—Standard Test Method for Rebound Number of Hardened Concrete, vol. 04.02. ASTM International (2013)

Caltrans: Test Method 550—Method of Test for Surface Abrasion Resistance of Concrete Specimens, Dec 2013

Caltrans: Test Method 551—Method of Test for the Suitability of Materials for Overlayment and Repair of Portland Cement Concrete Pavement and Structures, Aug 2012

Haworth, M.: Interstate 80 Donner Summit Shotblasting and Transil Application, pp. 1–4. Testing Report, Blastrac, Inc., Oklahoma City, Oklahoma (2011)

Nasvik, J.: Lithium Silicate Densifiers, pp. 1–5. Concrete Construction. On-line, Available: http://www.concreteconstruction.net/concrete-construction/lithium-silicate-densifiers.aspx (Dec 2008). Accessed in Aug 2017

Riemer, C., Pittenger, D., Gransberg, D.D.: preservation of concrete pavement using a modified silicon reactive lithium surface densifier over shotblasting: a life cycle cost analysis. 2012 Transportation Research Board Meeting. Washington, D.C., Jan 2012

Stokes, D.: Lithium ion penetration in the route 113 treatment site. Letter Report, pp. 1–2. FMC Corporation, Lithium Division, Bessemer City, North Carolina, July 28, 2010

An Optimized Data Interpretation for Marshall Flow and Stability Test

Byoung Hooi Cho[1], Jinwoo An[1], Heejung Youn[2], and Boo Hyun Nam[1(✉)]

[1] Department of Civil, Environmental, and Construction Engineering, University of Central Florida, 12800 Pegasus Drive, Suite 211, Orlando, FL 32816, USA
boohyun.nam@ucf.edu

[2] School of Urban and Civil Engineering, Hongik University, 94 Wausanro, Mapo-Gu, Seoul 04066, South Korea

Abstract. Due to its simple and quick manner in testing, Marshall Stability and Flow test method is still widely used over the world. Testing method provides useful information such as strength, deformation, and stiffness represented by Stability, Flow, and Martial Quotient (MQ) respectively. However, an initial error in load-displacement curves can generate significant misleading in those mechanical properties. This paper presents an optimized data processing method for data interpretation. Previous researches and specifications related to MQ are reviewed to identify limitations of the exiting data interpretation methods. Laboratory experiments were performed to verify the limitation of the current methods. This study illustrates that a non-resistive section of specimen causes an "initial delay" in the load-deformation curve, resulting in misleading of the mixture properties. Lastly, the automated algorithm to offset the initial delay and determine the best MQ was devised.

1 Introduction

The concepts of the Marshall Stability and Flow test were developed by Bruce Marshall (1939). Since 1948, the Marshall test has been enormously developed and improved by the U.S. Army Waterways Experiment Station (WES) for the asphalt mixture design criteria (USCOE 1948). Although the Marshall test is weighted toward empirical approach with limited theoretical background, the test has been widely used and adopted in many countries, especially in Europe, for the last 60 years. Currently many of those agencies are using the Marshall test procedure with their own modifications and interpretations of the results.

In the late 1980s, several researchers began to have more interest in the Marshall Quotient (MQ) which is the ratio of Stability to Flow. MQ represents stiffness characteristic of the testing material. Lees (1987) mentioned the term "Marshall Quotient" at first in his "range method" research paper (Lees 1987). He also suggested the MQ as a criteria of asphalt mix design and recommended different levels of MQ for pavement thickness design under diverse conditions of traffic volume as shown in Table 1.

New York State Department of Transportation (NYSDOT) and several other DOTs in USA has adopted the Marshall Quotient as an appraisal standard in their projects

Z. Hossain et al. (eds.), *Solving Pavement and Construction Materials Problems with Innovative and Cutting-edge Technologies*, Sustainable Civil Infrastructures,
https://doi.org/10.1007/978-3-319-95792-0_10

Table 1. Lees' proposed criteria of asphalt mix for pavement design

Specification			Commercial vehicles per lane per day		
			<1500	1500 ∼ 6000	>6000
Min. Marshall stability (kN)			2.2	3.3	6.7
Min. Marshall flow (mm)			2.0	2.0	2.0
Marshall Quotient (kN/mm)	Layer thickness (in.)	<2	0.6 ∼ 0.9	0.9 ∼ 1.2	1.4 ∼ 1.9
		2 ∼ 4	0.9 ∼ 1.2	1.2 ∼ 1.6	1.9 ∼ 2.5
		>4	1.2∼	1.6∼	2.5∼

(Gupta and Van Bramer 1989; Likitlersuang and Chompoorat 2016). Many counties have specified the Marshall Quotient as specification like as listed in Table 2. The specifications have their own criteria established through researches and developments related to the Marshall Quotient.

Marshall Quotient (*MQ*) could be simply calculated as the ratio of stability to flow from the Marshall test as following Eq. (1). Marshall stability (*S*) is the maximum magnitude of applied load on the test specimen and Marshall flow (*F*) is amount of deformation corresponding the maximum load value toward the loading direction.

$$MQ = S/F \tag{1}$$

where, MQ = Marshall Quotient (kN/mm)
S = Marshall stability (kN)
F = Marshall flow (mm).

Although this traditional interpretation method is quick and simple to use, the current interpolation method is often not accurate because it may not adequately reflect initial errors due to partial contact between specimen and the Marshall head. The initial errors could include a slightly misaligned installation of equipment and specimen and/or uneven surface condition of specimens. These undesired, but unavoidable, setups can cause "initial delay" in the load-deformation curve, resulting in errors that mislead the results such as overestimated flow, and underestimated MQ values consecutively.

Due to the initial delay have been universally occurred, ASTM D6927 developed concept to identify flow value in version of 05e1, which is deformation from the point where the projected tangent of the linear part of the curve intersects the x-axis to the point corresponding stability value (ASTM D6927-05e1 2005). However, the ASTM standard has not provided detail procedure, also been quite difficult to identify range of linear part on the curve in many cases. Moreover, the tangent line tends to be stiff, if the line is drawn at inflection point on the curve in sometimes.

In this paper, a series of Marshall stability and flow tests were performed to identify the initial errors, particularly investigating how much the error adversely affect the data interpretation and also how much raw data are reliable. Alternative data interpretation

Table 2. Specifications of Marshall Quotient by different countries

Country	Specification	Marshall Quotient (kN/mm)	
		Criteria	Value
Australia	WA specification 245 (AAPA 2000)	# of compaction	
		35 blow	>1.0
		50 blow	>1.7
		75 blow	>2.0
Nepal	Standard specifications for road and bridge works (Department of Roads 2015)	Type of bitumen	
		Viscosity grade	2.0 ~ 5.0
		Modified	2.5 ~ 5.0
Italy	ANAS specification (Celauro et al. 2010)	Surface course	>3.5
		Binder course	>3.5
		Base course	>3.0
Indonesia	DGA specification (Dardak et al. 1994)	# of 80 kN axles	
		Up to 1.0×10^6	1.0 ~ 4.0
		Up to 5.0×10^6	1.8 ~ 5.0
England (Surrey County)	Surrey design guide (SLGA 2002)	Type of mixture	
		Type A	0.6 ~ 1.1
		Type B and C	>1.1
Spain	Spanish general technical specifications for highway rehabilitation (Valdés et al. 2011)	–	<8.0

method was then devised to minimize initial error and to secure better reliability for the test results with more focusing on stiffness characteristic of the material itself.

2 Laboratory Test and Data Analysis

2.1 Marshall Stability and Flow Test

The Marshall stability and flow tests were performed under varied mix design of asphalt concrete. Cylindrical bituminous specimens of 4-in. diameter and 2.5-in. height were prepared by following ASTM D6926 (ASTM 2016). Also, all tests were conducted in three samples with the same mix design by following the standard procedure outlined in ASTM D6927-15 (ASTM 2015). Figures 1 and 2 illustrate the Marshall testing setup used in this study and conceptual diagram for assembly of Marshall head and specimen respectively.

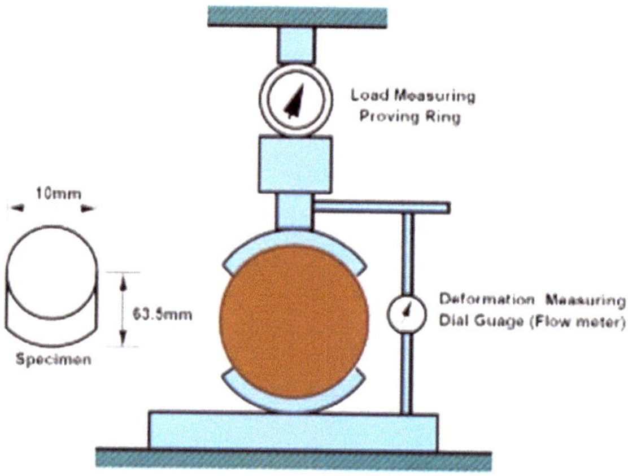

Fig. 1. A schematic diagram of Marshall testing machine

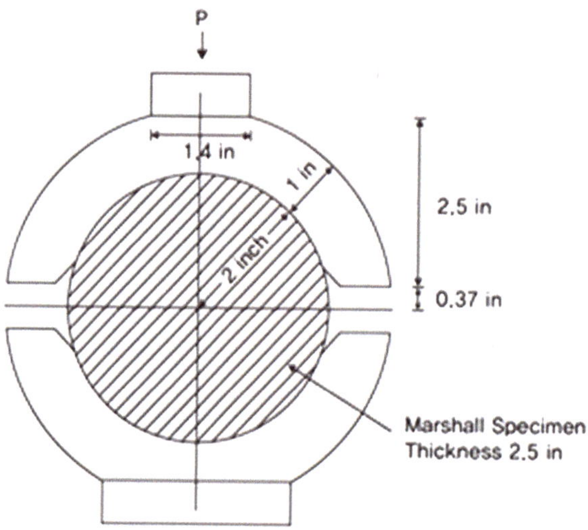

Fig. 2. Assembly of Marshall head and specimen

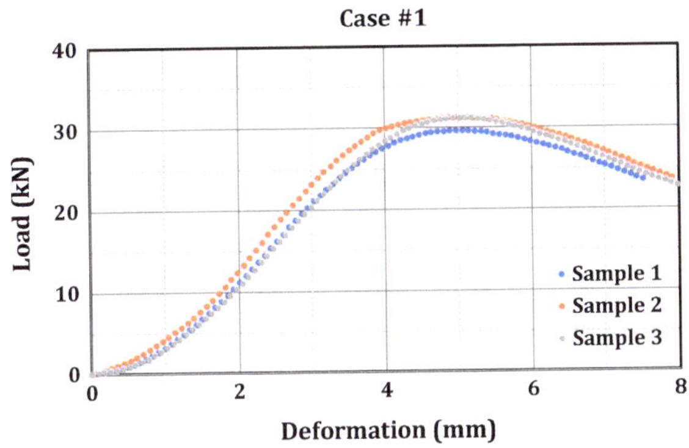

(a) Case of ideally performed the test

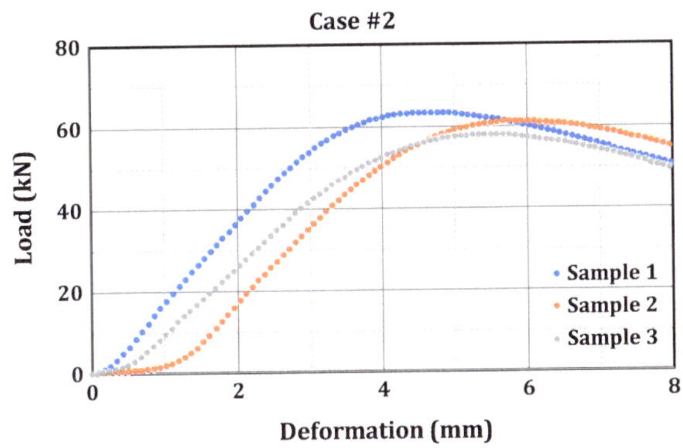

(b) Case with typical initial error occurred

Fig. 3. Cases of load-deformation graph from Marshall test

2.2 Data Analysis

The initial error has been detected in several test cases as expected (see Fig. 3b). This is identified by the initial delay in the load-deformation graph, which is estimated to be due to non-resistive sections resulting from subtly misaligned equipment installation or the specimen's uneven surface conditions etc. Figure 3. illustrates the load-deformation graphs of two cases: Case #1—ideal case without initial delay and Case #2—typical case with initial error from Marshall stability and flow test. Due to these graphs were obtained from tests for varied fiber modified mix design, the stabilities are larger than general Marshall mixtures.

As illustrated in the Fig. 3b, non-resistive sections were detected at the early stage of the test cycle. Particularly in Sample 2 (Orange dot-line), the load-deformation curve began to be plotted after about 1 mm shifted (delayed) from the starting point. It could be intuitively recognized that this initial delay has frequently occurred within ranging from 0 to 1.2 mm and often larger. It is assumed that the initial delay could cause the reliability issue in data interpretations.

Table 3. Comparison of statistical data for Case #1 and Case #2

Classification		Case #1	Case #2
Stability (kN)	Sample 1	29.53	63.23
	Sample 2	31.17	60.95
	Sample 3	31.17	57.97
	Average	30.62	60.72
	Standard deviation	0.95	2.64
	MCCOV	**3.09%**	**4.35%**
Flow (mm)	Sample 1	5.07	4.68
	Sample 2	5.03	5.83
	Sample 3	5.02	5.42
	Average	5.04	5.31
	Standard deviation	0.02	0.58
	MCCOV	**0.45%**	***10.98%***
Marshall Quotient (kN/mm)	Sample 1	5.83	13.51
	Sample 2	6.19	10.46
	Sample 3	6.20	10.69
	Average	6.07	11.55
	Standard deviation	0.21	1.70
	MCCOV	**3.51%**	***14.73%***

Averages, standard deviations and mean-centered coefficient of variation (cov) were computed for the stability and flow directly measured from the Marshall test. And the statistical values were compared to verify the differences between the ideally performed and initial delay occurred cases. In the same way, the statistical analysis was conducted for the Marshall Quotients calculated by the conventional method, which is representing stiffness characteristic of the asphalt concrete mixtures.

Table 3 contains all data not only from the Marshall tests but also from the statistical calculations. Mean-centered coefficient of variation (MCCOV) has represented the ratio about how much data have been dispersed (i.e. ratio of standard deviation to average). The statistical data indicates that Case #2 (with initial delay) is more dispersed then Case #1 (without initial delay).

3 Alternatives for Data Interpretation

3.1 Determination of Modified Marshall Quotient

Minimizing the effect of the initial delay plays a pivotal role in getting more reliable values of the material properties such as flow and Marshall Quotient. For this reason, an alternative method has been contrived, which could automatically offset the initial delay when flow and Marshall Quotient from the plotted load-deformation curve are computed. Figure 4 illustrates the algorithm to enable the automation of data processing to determine the modified flow(flow′) and modified Marshall Quotient (MQ′).

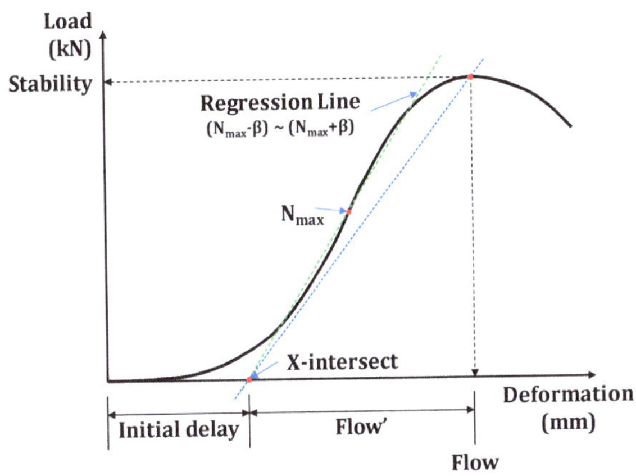

Fig. 4. Procedure of modified Marshall Quotient determination

The algorithm is composed of six steps as follow.

Step 1 Find the value of stability and flow using the traditional method.

Step 2 Find data point, N_{max}, having maximum slop$_{(N)}$. Here, the slop$_{(N)}$ could be calculated by Slop$_{(N)}$= (Load$_{(N)}$ − Load$_{(N-1)}$)/(Deformation$_{(N)}$ − Deformation$_{(N-1)}$).

Step 3 Draw linear regression line from data point ($N_{max} - \beta$) to ($N_{max}+ \beta$). Here, β is the number of data point from N_{max}. Also, least square curve fitting method is used to find the linear regression line (equation).
(For example, if the maximum slop is founded at data point #28 and β is selected as 5, the linear regression line is calculated from data point #23 to #33, which including total 11 data points.)

Step 4 Find X-intersect from the linear regression equation. This will be an amount of "initial delay" in this alternative method.

Step 5 Compute the value of modified flow(flow′) by flow′ = flow − initial delay. The influence of initial delay causing data misinterpretation is to be offset though this step.

Step 6 Compute the modified Marshall Quotient (MQ′) by MQ′ = stability/flow′. This MQ′ could not only reflect more impact of stiffness characteristic of the material itself, but also reduce the effect of the initial experimental error.

3.2 BETA (β) Selection

As defined before, β is the number of data point from N_{max}. Selection of appropriate β is crucial for the accurate data interpretation in the devised automated algorithm. If β is too small such as 0 or 1, a slop of the linear regression line would be too stiff (closer to ASTM method). This possibly induces overestimated stiffness properties of a material. On the other hand, once the β is selected too large like as more than 10, the stiffness of materials would be underestimated. Additionally, the non-resistive section (represented as initial delay) in the load-deformation graph could be included in the range used for regression analysis. This could drive distorted results of the stiffness characteristic as well. Moreover, a proper selection of β will produce consistent and undispersed slope values of the regression line from repetitive tests on the same mixture, although different initial delays are imbedded in the load-deformation graphs.

To find the optimum number of data point (β), mean-centered COV(MCCOV) were calculated and compared as changing β from 1 to 9, regarding as (1) slope of the linear regression line, (2) flow′ and (3) MQ′. Figure 5 presents typical cases of data trends for the computed MCCOV as changing thc β. The calculated MCCOVs have been consistent within a certain range of β (Red Boxes). Based on these results, the optimum value of β is determined as "5".

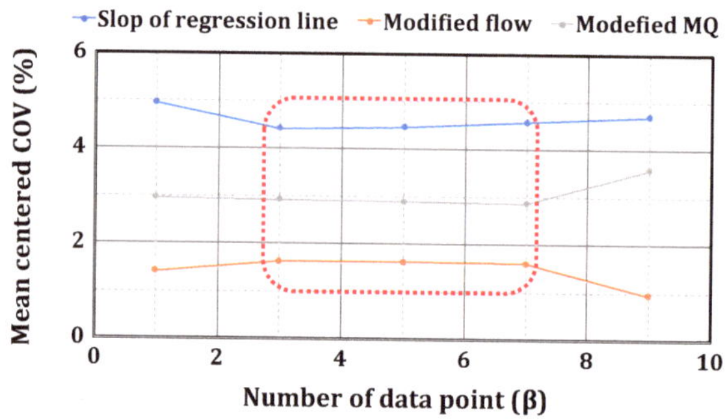

(a) Trend #1: Consistent MCCOV between β=*3 to 7*

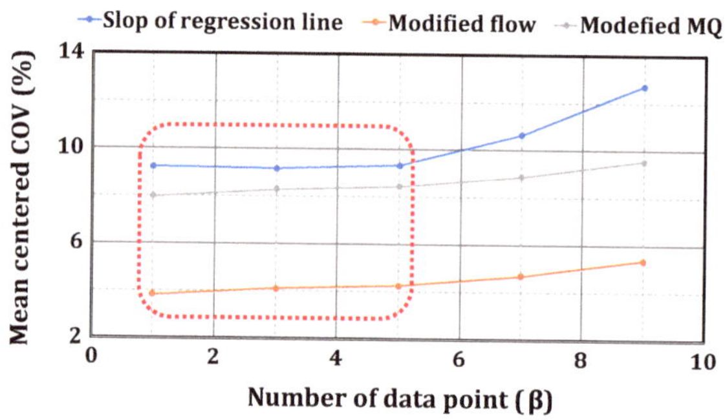

(b) Trend #2: Consistent MCCOV between β=*1 to 5*

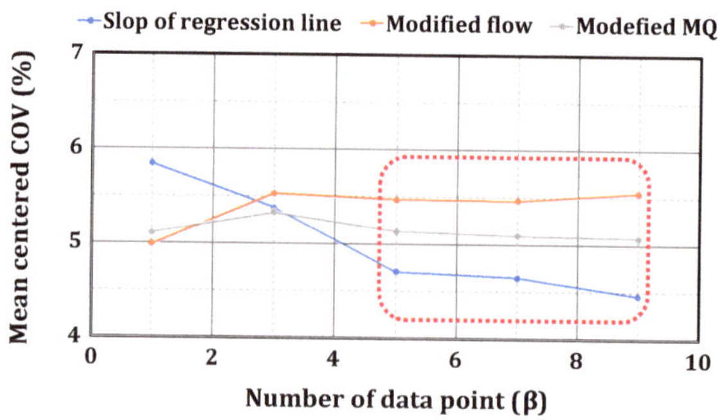

(c) Trend #3: Consistent MCCOV between β=*5 to 9*

Fig. 5. Typical cases of data trends of MCCOV as changing β

3.3 Data Analysis

The dataset of Case #1 and Case #2, which were previously handled in the Fig. 3 and Table 3, were re-analyzed implementing the devised interpretation algorithm. Since the alternative method has focused on diminishing effects of initial delay and reducing variances of interpreted values, it could be expected that effect of the new method might be greater in the Case #2 than in the Case #1.

As shown in Table 4, the devised data interpretation method in this study has improved the reliability of data interpretation in terms of the MCCOVs are decreased in Case #2 by more than 6% for both of flow and MQ. On the other hand, there was no significant improvement in Case #1, which is reasonable because Case #1 does not bear the initial delay.

Table 4. Comparison between traditional and the devised method

Classification		Case #1		Case #2	
Analysis method		Traditional	Devised	Traditional	Devised
Flow/Flow′[a] (mm)	Average	5.04	4.45	5.31	5.10
	Standard deviation	0.02	0.07	0.58	0.21
	MCCOV	**0.45%**	**1.60%**	***10.98%***	***4.20%***
MQ/MQ′[b] (kN/mm)	Average	6.07	6.88	11.55	11.93
	Standard deviation	0.21	0.20	1.70	1.00
	MCCOV	**3.51%**	**2.89%**	***14.73%***	***8.42%***

[a]Flow is from traditional method and flow′ is from the devised method
[b]MQ is from traditional method and MQ′ is from the devised method

3.4 Scenario Analysis

Scenario analysis was performed to verify how the devised data interpretation algorithm could effectively offset the initial error and reasonably calculate the MQ′ that better reflects the stiffness characteristics of the material itself.

Dataset for the scenario analysis was implemented by modifying the actual dataset as shifting initial point within a reasonable range of which initial delay could occur. As illustrated in the Fig. 6, the data conditions are like as followings;

(1) Three Marshall curves are plotted for three samples on same mix design.
(2) It is expected that samples have similar stiffness (comparable slop degree).
(3) Initial delays of each sample are set as about 1.0, 0.5 and 0.0 mm intuitively.

Table 5 shows the advantages of the automated interpretation method for the Marshall load-deformation curve. Initial delay about 1.0 mm (sample 1) has induced more than 20% of difference in flow value and more than 25% of difference in MQ value for the two different methods. On the other hand, a difference of less than 1% is displayed in sample 3, which does not bear an initial delay.

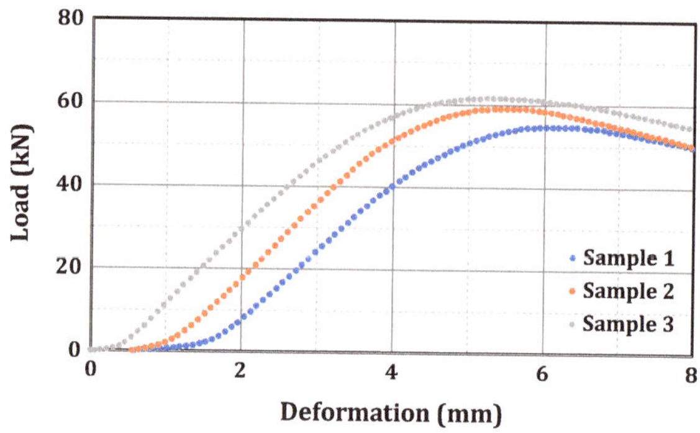

Fig. 6. Marshall load-deformation graph for scenario analysis

Table 5. Comparison of scenario data interpretation

Classification		Traditional method	Devised method	Difference (%)
Flow/Flow′[a] (mm)	Sample 1	6.19	4.95	**1.24 (20.13%)**
	Sample 2	5.59	4.83	**0.76 (13.61%)**
	Sample 3	5.40	5.35	**0.05 (0.87%)**
	Average	5.73	5.04	–
	Standard deviation	0.42	0.28	–
	MCCOV	**7.28%**	**5.46%**	–
MQ/MQ′[b] (kN/mm)	Sample 1	8.82	11.05	**2.23 (25.21%)**
	Sample 2	10.56	12.22	**1.66 (15.75%)**
	Sample 3	11.39	11.49	**0.10 (0.87%)**
	Average	10.26	11.59	–
	Standard deviation	1.31	0.60	–
	MCCOV	**12.77%**	**5.13%**	–

[a]Flow is from traditional method and flow′ is from the devised method
[b]MQ is from traditional method and MQ′ is from the devised method

Also, the devised method in this study has clearly shown improvement of reliability for data interpretation. It could be assumed that the Marshall Quotients of the three samples might be within small differences because they have closed degree of slops. As expected, the MQ′ by the automated algorithm have expressed more consistent values than those by the conventional data interpretation method. Moreover, MCCOV have been decreased by using the new method as previously confirmed. In other words, the variance of the dataset is reduced as effectively eliminating the initial error.

4 Conclusions

In this study, it was found that the conventional data interpretation method produces less reliable results. The traditional method does not offset the initial experimental error due to the initial delay when computing MQ which represents the material's stiffness characteristic. In addition, the MQ values from the method have shown a large range of data variation. Alternatively, an optimized algorithm of data interpretation was developed. The devised method could automatically eliminate disadvantages of the initial delay in the load-deformation curve. Consequently, the developed algorithm has provided more consistent results from a statistical point of view. Case studies have verified these improvements decreased more than 6% in the variation of the results compared to the traditional method. Moreover, it was identified that the initial delay of about 1.0 mm has induced more than 20% of difference in flow value and more than 25% of difference in MQ value between the two methods. On the other hand, a difference of less than 1% is displayed in the case not bearing the initial delay. From these results, it has been concluded that the newly devised algorithm could automatically reflect the initial error and reasonably compute the mechanical properties of materials.

References

ASTM: Standard Test Method for Marshall Stability and Flow of Bituminous Mixtures, ASTM International, D6927-05e1. West Conshohocken, PA (2005)

ASTM: Standard Test Method for Marshall Stability and Flow of Asphalt Mixtures, ASTM International, D6927-15. West Conshohocken, PA (2015)

ASTM: Standard Practice for Preparation of Asphalt Mixture Specimens Using Marshall Apparatus, ASTM International, D6926-16. West Conshohocken, PA (2016)

Australian Asphalt Pavement Association: Western Australia Specification 245 Asphalt Concrete, AUS-SPEC-2. WA, Australia (2000)

Celauro, C., Bernardo, C., Gabriele, B.: Production of innovative, recycled and high-performance asphalt for road pavements. Resour. Conserv. Recycl. **54**(6), 337–347 (2010)

Dardak, H., Dachian, A.T., Toole, T., Edwards, A.C.: Research on the specification and design of hot-mix asphalt surfacings in Indonesia, In: Transport Research Laboratory, Proceedings of the Fifth Annual Conference on Rood Engineering, Bondung, Indonesia (1994)

Department of Roads: Standard Specifications for Road and Bridge Works—2073. Ministry of Physical Infrastructure and Transport, Government of Nepal (2015)

Gupta, P.K., Van Bramer, T.F.: Implications of adopting a" range Method" for New York's marshall mix design (No. FHWA/NY/SR-89/96,). Engineering Research and Development Bureau, New York State Department of Transportation (1989)

Lees, G.: Asphalt mix design for optimum structural and tyre interaction purposes. In: Sixth International Conference, Structural Design of Asphalt Pavements, vol. I, Proceedings, University of Michigan, Ann Arbor, Michigan. Publication of: Michigan University, Ann Arbor, 13–17 July 1987

Likitlersuang, S., Chompoorat, T.: Laboratory investigation of the performances of cement and fly ash modified asphalt concrete mixtures. Int. J. Pavement Res. Technol. **9**(5), 337–344 (2016)

Surrey Local Government Association (SLGA): Surrey design: a strategic guide for quality built environments, Technical Appendix (2002)

U.S. Army Corps Engineers (USCOE): Investigation of the Design and Control of Asphalt Paving Mixture, Waterway Experiment Station, vols. 1 and 2 (1948)

Valdés, G., Pérez-Jiménez, F., Miró, R., Martínez, A., Botella, R.: Experimental study of recycled asphalt mixtures with high percentages of reclaimed asphalt pavement (RAP). Constr. Build. Mater. **25**(3), 1289–1297 (2011)

Use of Intelligent Compaction in Detecting and Remediating Under-Compacted Spots During Compaction of Asphalt Layers

Manik Barman[1(✉)], Syed Asif Imran[2], Moeen Nazari[3], Sesh Commuri[2], and Musharraf Zaman[3]

[1] Department of Civil Engineering, University of Minnesota Duluth, Duluth, USA
mbarman@d.umn.edu
[2] Nevada Advanced Autonomous Systems Innovation Center, University of Nevada, Reno, USA
[3] School of Civil Engineering and Environmental Science, The University of Oklahoma, Norman, USA

Abstract. Under-compaction of asphalt layers results in premature distresses like rutting, localized depressions and pot-holes. Over-compaction may crush the aggregates which can result in unstable asphalt mixes. It is therefore highly important to achieve the required air voids or relative density (6–8% air voids or 92–94% relative density). Real-time monitoring of the relative density can certainly be helpful in achieving the required relative density. The traditional quality control procedure, which involves collecting cores and conducting volumetric analysis on them, does not provide any measure of the air voids or relative density level during the compaction itself, thus under-compacted spots, if any, remain undetected. Intelligent compaction methods are able to continuously monitor the air voids or density of asphalt layers during the compaction process. The University of Oklahoma has developed an intelligent compaction analyzer (ICA). The ICA is based on the hypothesis that the vibratory roller and the underlying pavement form a coupled system whose response during compaction is influenced by the stiffness of the pavement layers. The ICA is capable of generating as-built maps providing information on coverage and quality of compaction of the compacted asphalt layers. This paper discusses the principle of ICA, and its application in measuring the density of asphalt layers. Results from one demonstration are included in this paper. The ICA measured densities were validated by comparing them with densities obtained from cores. It was found that the ICA measured densities and core densities correlated well with an R^2 between 0.85 and 0.93. Also, t-test conducted with the ICA-estimated densities and core densities verified that the difference between the above-mentioned two types of densities are insignificant at 95% confidence level. ICA was able to detect several under-compacted spots which were then remediated with additional roller passes. The application of the ICA certainly helped in achieving higher and uniform density throughout the test section.

Z. Hossain et al. (eds.), *Solving Pavement and Construction Materials Problems with Innovative and Cutting-edge Technologies*, Sustainable Civil Infrastructures,
https://doi.org/10.1007/978-3-319-95792-0_11

1 Introduction

Proper compaction of the asphalt layers is essential to achieve a long-lasting asphalt pavement. Under-compaction can leave a high percentage of air voids in the asphalt layers, making it susceptible to moisture infiltration, oxidation, rutting, and cracking (Arambula et al. 2007). On the other hand, over-compaction may result in (i) low air voids that could result in asphalt binder bleeding during hot weather conditions (U.S. Army Corps of Engineers 2000), and (ii) aggregate crushing that could lead to premature rutting. Therefore, achieving appropriate percentage of air voids (approximately 6–7% during the compaction) in the asphalt layer is very essential. The percentage of air voids of the compacted pavement layer is affected by the type of rollers, the number of roller passes, and the rolling patterns used in the compaction process (Scherocman and Dwight 2008). A good quality control (QC) mechanism involving real-time monitoring of the compaction during the construction process can identify and remediate inappropriate compaction, which can lead to an improved quality of the pavement (Beainy et al. 2012). Intelligent Compaction (IC) techniques are able to provide real-time information on the quality of compaction achieved during the construction process. Several Intelligent Compaction solutions that aim to provide real-time estimation of pavement quality during compaction are currently available in the market (Arasteh 2007; Briaud and Seo 2003; Rakowski 2008). The performance of these systems in guaranteeing the quality of compaction during construction of asphalt pavement has been widely studied (Chang et al. 2011, 2014). One of the limitations of these systems is that the compaction quality is reported in terms of a manufacturer specific roller measured value (RMV) and not in terms of design parameters such as density or stiffness of the mix (Barman et al. 2016; Imran et al. 2017). Additional time-consuming tests have to be performed in order to develop correlation values between RMV and design parameters such as density or stiffness (Chang et al. 2014).

Intelligent Compaction Analyzer (ICA) is a device developed by the researchers at The University of Oklahoma, Norman, OK, United States of America (USA). The ICA can be used for real-time estimation of pavement layer density and/ or stiffness during compaction (Beainy et al. 2012; Singh et al. 2011). In previous studies (Barman et al. 2016; Singh et al. 2011; Commuri 2010), the ICA was used to determine the quality of compaction of stabilized subgrade and asphalt layers during construction. The ICA was capable of generating as-built maps providing information on coverage and quality of compaction of the constructed pavement. While the use of the ICA in QC of the asphalt pavement construction was investigated, its use in improving the overall quality of compaction was not studied. The current study addresses the application of the ICA in QC as well as quality improvement (QI) of asphalt pavements. Field demonstration of the ICA verifying that ICA can be used to identify and remediate under-compacted spots is included in this paper. It was found that the ICA can be suitably used for QC as well QI of asphalt pavements.

2 Intelligent Compaction Analyzer (ICA)

Components: A picture of the ICA integrated roller is shown in Fig. 1a. The ICA consists of a uniaxial accelerometer for collecting the roller vibrations, a roller-mounted rugged tablet computer for extracting the features from collected roller vibrations, an ANN-based classifier, a user interface, and a display system. A GPS receiver is mounted on the roof of the roller and georeferenced with respect to the axle of the roller drum. Generally, a Trimble ProXT series GPS receiver with position accuracy at the submeter level is used in the ICA system. A smooth drum roller is used for compaction. The Summit Instruments 13200C uniaxial accelerometer is installed in the axle of the

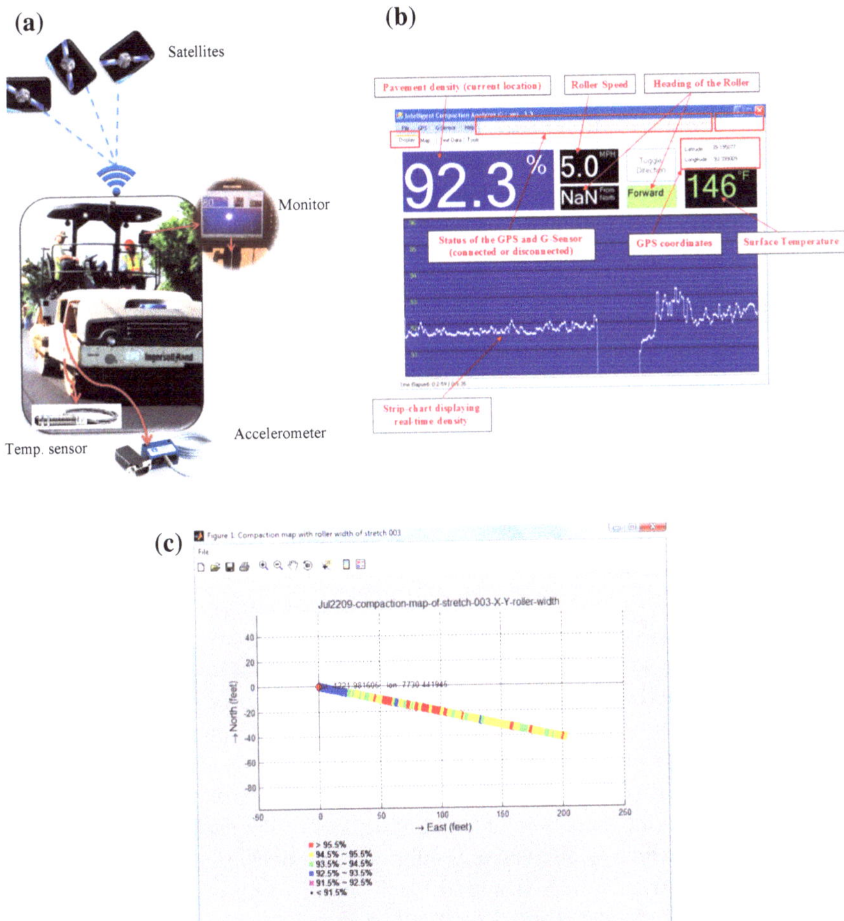

Fig. 1. **a** Major components of ICA. **b** Screen-shot showing the process information during compaction of asphalt layers (Commuri 2010). **c** Screen-shot showing the as-built compaction strip chart that can be seen real-time by the roller operator during the ICA compaction (Commuri 2010)

smooth drum roller. This uniaxial accelerometer has a full-scale range of ±10 g and typical amplitude linearity remaining within 0.2% for the entire range.

Working Principle: The ICA is based on the hypothesis that the vibratory roller and the underlying pavement form a coupled system, and the roller vibrations during compaction are characteristic of the stiffness of the underlying material (Commuri et al. 2012; Commuri and Zaman 2008). Any variations in stiffness in the underlying pavement layer cause variations in the roller's vibration pattern. These variations can be captured and analyzed to estimate the stiffness of the layer being compacted. The spectral analyses are performed on the vibrations of the roller drum. The power at each frequency level is computed, and pertinent features are extracted from the frequency spectrum for vibrations corresponding to different levels of compaction. An Artificial Neural Network (ANN) is then used to classify the vibration patterns in real-time and correlated with appropriate stiffness levels. These stiffness levels are then converted to asphalt layer density in the calibration module of the ICA. The ICA-display then combines geographic locations of spots, determined by the on-board GPS, and asphalt layer density values to present compaction data in real-time to the roller operator. A complete description of the ICA's working principle can be found at the following publications: Barman et al. (2016), Imran et al. (2017), Beainy et al. (2014), Commuri and Zaman (2008), and ICA User Manual (Commuri 2010).

ICA Calibration: During field compaction, the ICA is first trained to recognize the vibration patterns; the trained ICA is then calibrated to convert these patterns into a numerical value indicative of the density of the layers being compacted. The ICA provides as-built maps showing process information such as number of roller passes, roller path, GPS coordinates of the roller (see Fig. 1b), and a color coded as-built map showing the asphalt layer density in real-time (see Fig. 1c). Access to compaction quality in real-time enables the roller operator to identify and remediate any under-compacted areas on the asphalt layer being compacted and thereby improve the quality of construction.

3 ICA Demonstration

The conventional quality control of asphalt layer compaction involves extracting cores from randomly selected locations. Densities of these cores are then measured to determine the achieved compaction level. This practice however does not provide any real-time quality control and covers only a fraction of the compacted area. In ICA compaction procedure, the compaction quality is continuously monitored on the entire compacted area by measuring the relative density (%) of asphalt mix. The ratio of the compacted density to the theoretical maximum density of the asphalt mix is known as the relative density which is alternatively referred to as density (%) in this paper.

In the current study, the ICA was used for the compaction of the base and surface layers of a rural road (Name: Acme Road) at Shawnee, Oklahoma, USA. The asphalt mixture design information for both the layers is provided in Table 1. Both of the mixes were prepared with PG 64-22 binder. The nominal maximum aggregate size (NMAS) for the base- and surface- layer was 25.4 mm (1 in.) and 12.7 mm (1/2 in.), respectively. Both the mixes contained a significant percentage of reclaimed asphalt

Table 1. Properties of asphalt mixtures in Acme Road Project

Parameters	Base layer	Surface layer
Nominal maximum aggregate size (mm)	25.4	12.5
Los Angeles abrasion (%)	23.7	23.4
Effective specific gravity of aggregates	2.707	2.693
Type of asphalt binder	PG 64-22	
Proportion of RAP in the asphalt mix (%)	25	35
Specific gravity of asphalt binder	1.010	
Asphalt binder content	4.0% (total); 3.0% (virgin)	4.7% (total); 3.5% (virgin)
Maximum theoretical specific gravity of asphalt mix	2.535	2.495
Voids in mineral aggregates (%)	13.6	15.2

pavement (RAP), 25% in the base and 35% in the surface layer. Figure 2a, b show photographs of the ICA integrated roller during the compaction of the base and surface layers, respectively.

(a) (b)

Fig. 2. ICA compaction in progress, in Acme Road: **a** Base layer, **b** surface layer

After the installation of the ICA hardware (see Fig. 1a) on the roller, the ICA was calibrated on a 9.14-m (30-ft) long calibration stretch, separately for base and surface layers. The paving contractor was requested to pave the calibration stretch first. Vibration and GPS data were collected on several roller passes during the compaction of the calibration stretch. The vibration data were then used to train the ICA to recognize the power features in the vibratory signal. A preliminary calibration of the ICA was performed considering an approximate laydown density (88.1% and 86.5 for base and surface, respectively) and a maximum final density based on the mix design information obtained from the contractor. In order to adjust the calibration parameters, three cores' locations were marked on the calibration stretch after the compaction was performed. Cores were extracted from the marked locations and their densities were determined later in the laboratory according to the AASHTO T-166 test method.

Figure 3 shows the locations of the cores on the base and surface layers. Final calibration of the ICA was performed by comparing the ICA-estimated densities at the calibration stretches (on base and surface layers) with the density of the cores collected from the calibration stretches on the base and surface layers. Further details of the calibration procedure can be found in the ICA User Manual (Commuri 2010).

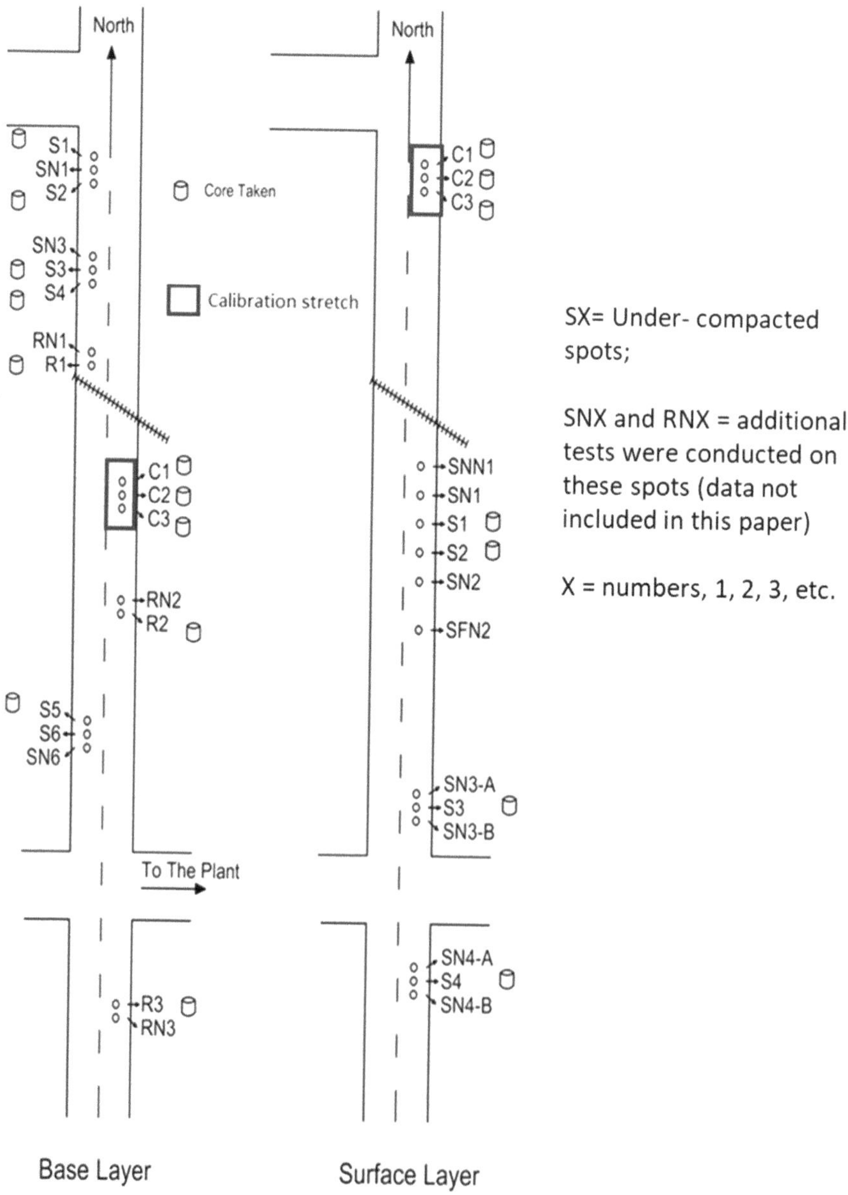

Fig. 3. Location of test points on base and surface layers of Acme Road

After the preliminary calibration of the ICA was performed (meaning that the ICA could estimate the approximate relative density of the asphalt layer being compacted), the compaction was performed on the remaining stretch of the project. During this process, roller vibrations, spatial locations (GPS coordinates) of the roller and surface temperature of the asphalt layer were collected in addition to collecting the ICA estimated relative densities (or ICA-estimated densities) throughout the project stretch. It shall be noted that ICA-estimated densities were later on adjusted based on the final calibration of the ICA.

4 Results and Discussions

4.1 Identification and Remediation of Under-Compacted Regions

The performance of the ICA in identifying and remediating the under-compacted regions was tested on both the base and surface layers of this project. Under-compacted-spots were identified at multiple locations during the compaction of the base and surface layers by monitoring the density as-built map in real-time. Figure 4 shows the locations of the under-compacted spots. Additional roller passes were provided to improve the density in those locations, wherever possible. ICA measurements were recorded both before and after the remedial passes.

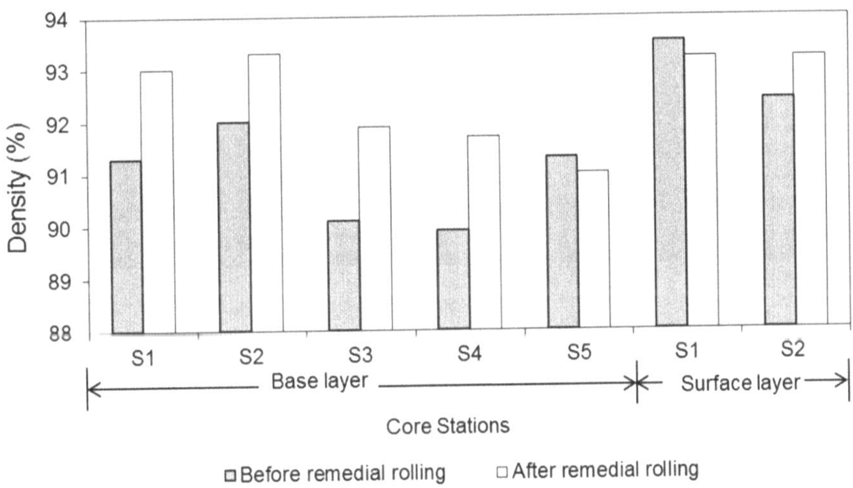

Fig. 4. Improvement in ICA-estimated density with remedial roller passes at selected under-compacted spots

Base Layer: In the base layer, six under-compacted spots (S1–S6) were identified. Three remedial roller passes were applied on first 4 spots (S1–S4), whereas only one additional pass could be provided on S5. The roller operator could not provide additional passes on S6 because of time constraint. It may be noted here that the contract of this construction work did not have any scope for ICA compaction, so additional roller passes could not be applied on all the under-compacted spots. Figure 5 presents a comparison of the ICA-estimated density measured before and after the remedial passes. It can be seen that the density at the four spots (S1–S4) was increased with the three remedial passes. The density at spot S5 could not be increased because only one additional pass was provided at this spot compared to three remedial passes on the other spots. The average density of the five spots could be improved from 90.9 to 92.2%.

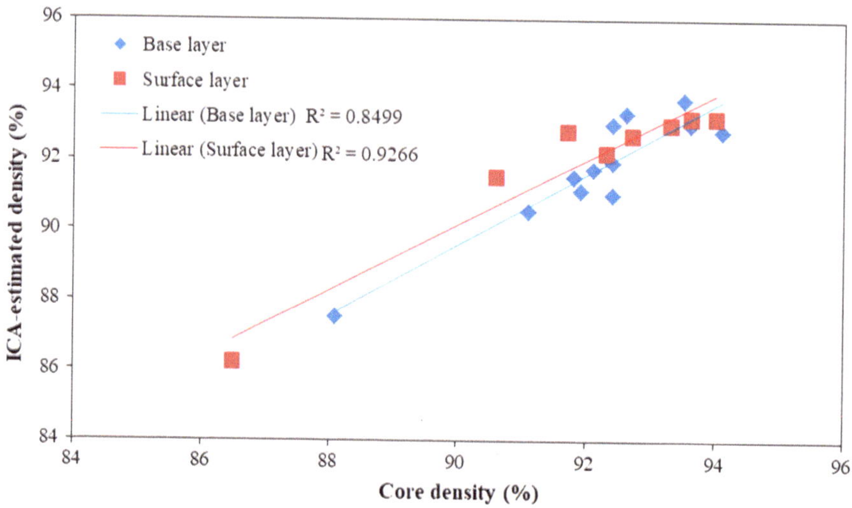

Fig. 5. Correlation between ICA-estimated densities and core densities

Surface Layer: In the surface layer, four under-compacted regions (S1–S4) were identified. However, remedial passes could only be applied on first two regions (S1 and S2) because of time constraint. As shown in Fig. 5, the density was found to increase in one spot (S2), while it remained almost the same in the other spot (S1). The average density improved from 93 to 93.2% for the surface layer.

4.2 ICA-Estimated Density Versus Core Density

In order to validate the ICA-estimated densities, additional cores were extracted at various locations as shown in Fig. 4. A total of eleven cores (including 3 cores at the calibration stretch) were extracted from the base layer and seven cores were extracted from the surface layer. Volumetric analyses were performed on all these cores. The core densities varied between 91.05 and 94.09% in the base layer, whereas it varied between 90.64 and 94.05% in the surface layer. The lay down density in the base layer was 88.10% and 86.5% for the base and surface layers, respectively. The correlations between the core densities and the corresponding ICA-estimated densities for the base and surface layers are given in Fig. 5. The correlations are quite reasonable for both the base (R^2 = 0. 85) and surface (R^2 = 0.93) layers. From these correlations, it can be concluded that the ICA could measure the density with a reasonable accuracy.

Additionally, a ‘paired two tailed unequal variance ‘t-test’ was conducted at a 95% confidence level to test the significance of the deference between the ICA-estimated densities and core densities. Table 2 presents the ‘t-test’ results. It was found that the *p*-values (0.53 for base and 0.99 for surface) for both of the base and surface layers were significantly larger than the critical value, 0.05. Also, the t-stats for both of the base and surface (0.63 for base and 0.01 for surface) are lower than the t-critical (2.07 for base and 2.14 for surface). These indicate that the difference between the ICA-estimated densities and core densities were not significantly different at a 95% confidence level.

Table 2. t-test results for comparing the ICA-estimated densities and core densities

Parameters	Base layer		Surface layer	
	ICA-estimated density	Core density	ICA-estimated density	Core density
Mean	92.17	91.75	91.84	91.85
Variance	2.35	2.83	5.85	5.54
Observations	12	12	8	8
Hypothesized mean difference	0		0	
Degree of freedom	22		14	
t stat	0.63		−0.01	
P(T ≤ t) two-tail	*0.53*		*0.99*	
t critical two-tail	2.07		2.14	

5 Conclusions

This paper presents a brief discussion on the principle, components and application of the Intelligent Compaction Analyzer (ICA), developed at The University of Oklahoma As a demonstration, the ICA compaction was performed on the base and surface layers in one rural road in Oklahoma, USA to verify its applicability in quality control (QC) and quality improvement (QI) of the asphalt pavement during construction.

The ICA could provide real-time information about the level of compaction achieved in terms of relative density. The relative density is the ratio of the density at any location to the maximum theoretical density. The relative density was monitored throughout the compaction process. Under-compacted regions were identified during this process. Additional remedial passes were applied to improve the level of compaction on the identified under-compacted regions. The average density of the base layer in the demonstration project could be increased from 90.9 to 92.2%; whereas, for the surface layer, average density could be increased from 93 to 93.2%.

The ICA-estimated densities were validated by comparing them with the relative density of cores extracted from selected locations on the compacted asphalt layer. The coefficient of determination (R^2) for the correlation between the ICA-estimated densities and core densities were obtained between 0.85 and 0.93. Also, t-test conducted with the ICA-estimated densities and core densities verified that the difference between the above-mentioned two types of densities are insignificant at 95% confidence level. Based on this observation, it can be concluded that the ICA can be helpful in identifying and remediating any under-compacted regions in asphalt layers. However, in order to successfully implement the ICA based compaction, additional studies are recommended. Also, it is recommended that Intelligent Compaction be considered as a requirement in the bidding of the work for a very comprehensive research study.

Acknowledgements. The field demonstration presented in this report would not have been possible without the unparalleled support of Oklahoma Department of Transportation (ODOT), Haskell Lemon Construction Company (HLCC), Oklahoma City, Oklahoma and Silver Star Construction Company, Moore, Oklahoma. Their partnership with the University of Oklahoma has been critical for the success of this project.

References

AASHTO: Bulk Specific Gravity of Compacted Bituminous Mixtures Using Saturated Surface-Dry Specimens. AASHTO T-166, Washington, DC (2007)

Arasteh, M.: Innovations in compaction control and testing. In Intelligent Compaction 2007 Construction Conference, Federal Highway Administration, Bishmark, ND (2007)

Barman, M., Nazari, M., Imran, S.A., Commuri, S., Zaman, M.: Quality improvement of subgrade through intelligent compaction. Transp. Res. Rec. J. Transp. Res Board **2579**, 59–69 (2016). https://doi.org/10.3141/2579-07

Beainy, F., Commuri, S., Zaman, M.: Quality assurance of hot mix asphalt pavements using the intelligent asphalt compaction analyzer. J. Constr. Eng. Manag., 178–187 (2012). https://doi.org/10.1061/(asce)co.1943-7862.0000420

Beainy, F., Commuri, S., Zaman, M.: Dynamical response of vibratory rollers during the compaction of asphalt pavements. J. Eng. Mech., 04014039 (2014). https://doi.org/10.1061/(asce)em.1943-7889.0000730

Briaud, J.L., Seo, J.: Intelligent Compaction: Overview and Research Needs. Texas A&M University, College Station, TX (2003)

Chang, G., Xu, Q., Rutledge, J., Horan, R., Michael, L., White, D., Vennapusa, P.: Accelerated Implementation of Intelligent Compaction Technology for Embankment Subgrade Soils, Aggregate Base, and Asphalt Pavement Materials. FHWA-IF-12-002. Federal Highway Administration, Washington DC (2011)

Chang, G., Xu, Q., Rutledge, J., Garber, S.: A Study on Intelligent Compaction and In-Place Density. FHWA-HIF-14-017, Federal Highway Administration, Washington, DC (2014)

Commuri, S.: Intelligent Compaction User Manual, Version 2. University of Oklahoma, OK (2010)

Commuri, S., Zaman, M.: A novel neural network-based asphalt compaction analyzer. Int. J. Pavement Eng. **9**(3), 177–188 (2008)

Commuri, S., Mai, A., Zaman, M.: Calibration Procedures for the Intelligent Asphalt Compaction Analyzer. ASTM. J. Test. Eval. **37**(5), 454–462 (2009)

Commuri, S., Mai, A.T., Zaman, M.: Neural network-based intelligent compaction analyzer for estimating compaction quality of hot asphalt mixes. J. Constr. Eng. Manag. **137**(9), 634–644 (2012)

Imran, S.A., Commuri, S., Barman, M., Zaman, M., Beainy, F.: Modeling the dynamics of asphalt–roller interaction during compaction. J. Constr. Eng. Manag. (2017). © ASCE, ISSN: 0733-9364

Rakowski, S.: Intelligent compaction CCV IC. In: Transportation Pooled Fund Intelligent Compaction Systems Initial-Task Working Group Meeting, Sakai America, Adairsville, GA (2008)

Scherocman, J.: Compaction hot-mix asphalt pavements. Part I: Roads and bridges (2000). http://www.roadsbridges.com/compacting-hot-mix-asphalt-pavements-part-i (Apr. 6, 2016)

Singh, D., Beainy, F.N., Mai, A.T., Commuri, S., Zaman, M.: In-situ assessment of stiffness during the construction of HMA pavements. Int. J. Pavement Res. Technol. **4**(3), 131–139 (2011)

Design and Construct Contracts for Airport Asphalt Resurfacing

Greg White(✉)

Airport Pavement Research Program, University of the Sunshine Coast, Sippy Downs, QLD, Australia
gwhite2@usc.edu.au

Abstract. Traditionally, airport asphalt resurfacing works were delivered using a construct-only contract, requiring compliance with a design provided by the client, usually prepared by a design consultant. However, a number of generally compliant airport asphalt surfaces have failed to performed as required. Consequently, there is a desire to make asphalt contractors more responsible for the performance of airport asphalt surface that they have constructed. This paper describes the recent use of design and construct contracts for the delivery of airport asphalt resurfacing works in Australia. It is recommended that suitably sized and scoped airport asphalt resurfacing project consider a design and construct approach in the future. However, a robust preliminary design, nominated quantities for tendering and appropriate risk apportionment are all critical to project success.

1 Introduction

Traditionally, airport asphalt resurfacing works were delivered using a construct-only contract, requiring compliance with a design provided by the client, usually prepared by a design consultant (Rodway 2016). However, a number of generally compliant airport asphalt surfaces have failed to performed as required (White 2016a). Significant modes of distress include top–down cracking (Fig. 1), shear creep (Fig. 2) and groove closure (Fig. 3). Consequently, there is a desire to make asphalt contractors more responsible for the performance of airport asphalt surface that they have constructed. This paper describes the recent use of design and construct contracts for the delivery of airport asphalt resurfacing works in Australia. Background information covers airport pavements, runway resurfacing works and performance-based specification of asphalt. The drivers for transitioning to design and construction contracts are then outlines before the practical tender and contractual requirements are detailed.

2 Background

2.1 Airports and Pavements

Many Australian airports were developed by the Commonwealth Government in preparation for WWII or in reaction to the subsequent cold war period. At that time, the

Z. Hossain et al. (eds.), *Solving Pavement and Construction Materials Problems with Innovative and Cutting-edge Technologies*, Sustainable Civil Infrastructures,
https://doi.org/10.1007/978-3-319-95792-0_12

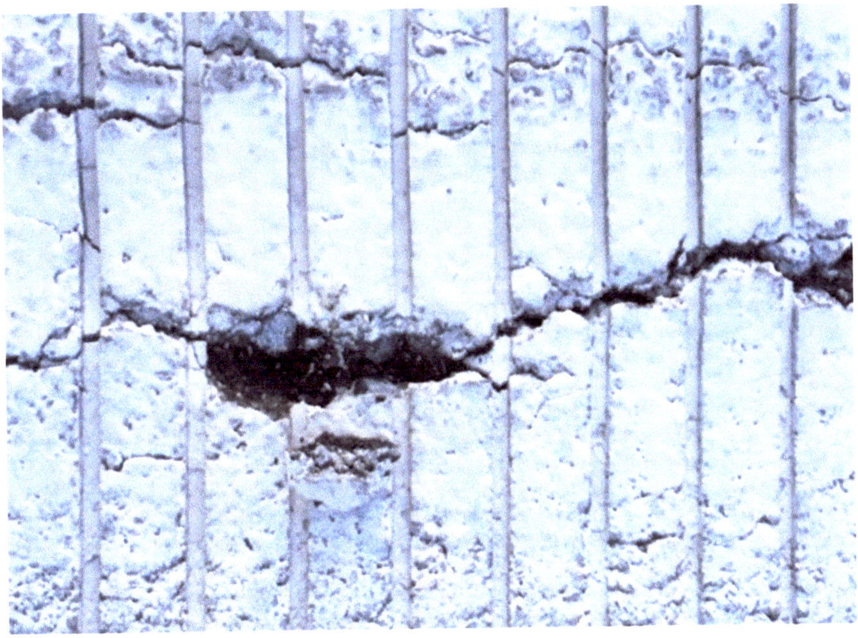

Fig. 1. Runway surface top–down cracking

Fig. 2. Runway surface shear creep

Fig. 3. Runway surface groove closure

Government set design and construction policies and standards, resulting in a nationally consistent approach, supported by Government-funded research and development for the upkeep of technology and practice. However, in the 1990s, the Commonwealth Government privatised the larger airports and transitioned the ownership of smaller and regional airports to local Government authorities. At the same time, the centralised team undertaking airport pavement research and development was disbanded, along with the centralised development and maintenance of policies, standards and guidance relating to airport pavement technology. Consequently, the privatised airports became the design authorities for their own airport pavement infrastructure. However, research and development ceased, best practice was not maintained and airport pavement technology stagnated.

2.2 Runway Resurfacing

Since the 1960s, Australia has generally followed the US Army Corps of Engineers (the Corps) approach to the design, construction and specification of airport pavements, including asphalt runway surfacing. Like the Corps, Australian airport asphalt has traditionally been designed to be densely graded with a high bituminous binder content. Polymer modified binders replaced conventional bitumen as the preferred asphalt binder around the year 2000, primarily for reduced asphalt temperature susceptibility and increased moisture damage resistance. Since that time, increases in aircraft tyre pressure and wheel load (White 2017a) in combination with reduced bitumen consistency (White 2016b) have resulted in examples of premature surface distress (White 2017b). In some cases, the distress has been severe and required significant and premature surface replacement. Consequently, some asphalt producers have developed proprietary polymer modified binders for reduced performance risk (White and Embleton 2015; Emery et al. 2015; Jansz 2017). In parallel, airports have expressed an increasing desire for a performance-based airport asphalt specification (White 2017c).

Runway resurfacing works are intense and high pressure. In Australia, almost all airports have only one main runway. This requires the majority of runway resurfacing projects to be completed in short night shifts with the runway returned to operational condition each morning. In practice, this requires a high production capacity to complete as much work as practicable in each work period, significant redundancy to reduce the risk of the runway not being serviceable at the end of each work period and temporary asphalt ramps to be constructed between the new and original surface.

2.3 Performance-Based Asphalt Specification

A performance-based asphalt specification for airport surfacing was developed on behalf of Australian airports, in conjunction with designers and asphalt producers (White 2017d). The primary aims of the specification are to allow the incorporation of proprietary binders, focus mixture design on asphalt properties that are indicative of asphalt field performance and to allow asphalt producers to be more responsible for asphalt field performance. A secondary outcome is to increase appropriate use of recycled materials. The performance-based asphalt specification achieves these aims by:

- Retaining the traditional dense graded asphalt volumetrics.
- Focusing on laboratory tests for deformation, fatigue and moisture damage resistance during mixture design.
- Allowing the mixture designer to select the bituminous binder.
- Preferring outcomes over methods for construction requirements.
- Retaining the traditional volumetric and Marshall properties for monitoring production consistency.

3 The Desire

Adoption of a performance-based asphalt specification is conveniently implemented in a design and construct contractual environment. However, there are other factors that also contribute to the desire for a design and construct approach to airport asphalt surfacing work.

3.1 Clear Responsibility

Under a traditional construct-only contract, the constructor's responsibility is appropriately limited to compliance with the design documentation, including the material specifications. Many contracts also include fitness for purposes obligations. However, the validity of requiring fitness for purposes from the contractor is questioned when a fully detailed and specified design must be followed.

Because airport resurfacing works are generally intense and relative short in duration the level of construction scrutiny is higher than for other civil construction works. As a result, forensic review of the quality records rarely explains the presence of significant distress because any evident deficiencies are identified and addressed during the execution of the works. Consequently, the quality records usually indicate that the works generally comply with the design, implying that the distress is symptomatic of poor performing, but compliant, airport asphalt surfacing.

When a compliant asphalt surfaces does not perform as expected, the design is often questioned. However, the design is usually determined to be consistent with industry practice. Consequently, the airport is left with an underperforming surface that is compliant with an appropriate design. Understandably, the airport is confused and questions why they are left to fund the cost of rectification when they already paid for adequate design and construction services. This situation has resulted in many airports preferring a single organisation to be fully responsible for the design and the construction of runway resurfacing works. A design and construct approach achieves this desire.

3.2 Efficient Design

Australian airports often receive grant funding from the State or Commonwealth Governments for airport resurfacing works. Although the funding is critical for regional airports, it is often provided with strict timeframes that place pressure on the design and construction program. One way to reduce this pressure is to undertake the construction preliminaries in parallel with the finalising the design. The construction preliminaries include work compound development, asphalt plant mobilisation and quality document preparation. This period is usually adequate to allow the design to be completed, including the asphalt mixture design and finalisation of the preliminary geometric design. Furthermore, requiring the contractor to finalise the design after the contract is tendered and awarded, the design can be better adjusted to meet the project budget and designs are less likely to present constructability challenges that are sometimes introduced by consultant prepared designs. This efficient approach to design has assisted to alleviate challenges for many runway resurfacing works, which is the desire of all airports. However, as will be discussed later, this approach relies on a robust basis for tendering the works before the design is finalised.

3.3 Innovative Solutions

As detailed above, the Australian Commonwealth Government ceased to maintain airport pavement technology and practice when the major airports were privatised in the 1990s. Since that time, many innovations have been developed overseas or in the Australian roading industry. In general, these innovations have been resisted by airports and designers due to a lack of demonstrable ‘local airport use and performance’. In many cases, potential solutions for improved performance, reduced cost or increased sustainability have been developed by the asphalt producers that execute both road and airport pavement resurfacing works. Procuring runway resurfacing through a design and construct approach reduces the resistance to introducing innovations by transferring liabilities from the risk-averse designer consultants to the contractors, who are better placed to manage and accept innovation risk.

Highlighting the general reluctance of designers to embrace innovation is not intended to be critical of the design consultants. Rather, it is simply an unfortunate circumstance. Designers are usually engaged through contracts with significant professional liabilities, discouraging the adoption or approval of solutions that are not an established element of historical local practice. Furthermore, designers are usually engaged to produce project-level documentation, not to undertake research and development for innovation advancement. The current construct-only procurement system is based on the premise that an independent third party maintains the state of the art, as occurs in many other countries. However, in Australia, this has not occurred in a systematic manner since the 1990s.

In the absence of Government coordinated research and development, construction contractors are better placed than designers to develop and implement innovations. Australian runway resurfacing examples of contractor-led innovations include proprietary products such as asphalt preservation treatments for improved surface life cycles (White and Thompson 2016), proprietary polymer modified binders for improved asphalt performance (White and Embleton 2015; Emery et al. 2015; Jansz 2017), the introduction of warm-mixed asphalt (White 2015) and the adoption of foamed bitumen stabilisation for expedient pavement reconstruction (White 2017e). Contractors are more willing to accept liability for the performance of their innovative solutions, but appropriately seek to protect the intellectual property associated with developed solution, in order to realise a return on their investment.

4 Requirements

The significant benefit associated with a design and construct approach to airport runway resurfacing is only realised if the works are procured appropriately. The over-riding requirement is to limit the contractor's liability to elements within their control. Otherwise, the contractor will perceive and value the risk. This contrasts the intent of many standard forms of contract that attribute the risk of all unknowns to the contractor. Further, the appropriate selection of projects for design and construct delivery is critical to success, as has been demonstrated internationally (Diefenderfer and Bryant 2006). With regard to runway resurfacing, the existing pavement must be in sound condition or its upgrade included in the preliminary design and nominated quantities. Where such a structural upgrade is required, either the contractor shoulder be responsible for validation of the entire pavement upgrade or the airport must determine the pavement upgrade requirement and nominate minimum asphalt resurfacing thicknesses to achieve the required structural capacity and the contractor relieved of this responsibility.

Once a suitable project is identified, attention must be given to the tender, contractual and post-completion requirements to maximise the benefit of the design and construct approach.

4.1 Preliminary Design

A preliminary design is critical for tendering runway resurfacing works under a design and construct contract. The preliminary design instills confidence in tenderers that the project is realistic, as well as ensuring the budget and scope are robust.

By their nature, resurfacing works are volume driven. Therefore, a realistic budget relies on the necessary volume of asphalt to be determined with confidence. The volume of asphalt is defined by the area to be resurfaced, the minimum thickness for structural requirements and the additional volume required for shape correction. It is important that the preliminary design consider all these aspects.

Geotechnical investigations and structural analysis is required to determine the minimum asphalt thickness required for structural capacity. Because a 10% error in the overall pavement structure thickness is typically equivalent to 100% or more of the resurfacing thickness, it is essential that the structural capacity be determined in a robust way. Unless some additional structural improvement is required under the same contract, for example a granular overlay or base course stabilisation, it is recommended that the airport remains completely responsible for the structural evaluation. The structural investigation and general scope of work will also determine the area of the pavement to be resurfaced, including or excluding shoulders, as well as the necessity to reshape the grass flanks.

The volume of asphalt required for shape correction is determined by a preliminary geometric design. This requires full engineering survey to be conducted on a grid consistent with the construction set-out. A three-point (shoulder, centreline and shoulder) survey and design is not adequate for a 30 or 45 m wide runway. Some airports have previously determined their budget by simply multiplying a nominal asphalt thickness by the area of their runway. This approach is possible, but the nominal thickness must take into account the likely additional average thickness required for shape correction, which is better determined from a preliminary geometric design.

4.2 Nominated Tender Quantities

Runway resurfacing is one area when airports get what they pay for reflecting the high influence the volume of asphalt has on the cost of the works. Tendering contractors should not be required to determine the volume of asphalt on which the tender is based. Rather, all tenders should be based on the same volume of asphalt and other volume-critical elements of work. The alternate would result in contractors determining whether the minimum asphalt thickness should be 40, 50 or 60 mm, which is not desirable because airports should control the volume of asphalt they eventually procure.

Nominated tender quantities are best derived from the preliminary design solution. Once the tender is awarded and the design is finalised, the tendered price must be reconciled to take account of the differences in quantities nominated in the tender and based on the accepted design solution. The reconciliation process emphasises the importance of a robust preliminary design and realistic quantities so as to minimise this change.

4.3 Tender Protocols

The tender must be based on the nominated quantities and the preliminary design solution. Where the performance-based asphalt specification is used, it is appropriate to require a draft asphalt mixture design in the tender submission. This approach allows the proposed asphalt mixture to be evaluated as a non-price element of the value-for-money proposition.

Due to the time required to undertake an asphalt mixture design, including the performance testing, a longer tender period is required, with eight weeks recommended. Furthermore, the cost of undertaking an asphalt mixture design during the tender period is significant. It is not reasonable to ask an unrestricted number of tenderers to incur this cost through an open tender process. Therefore, an expression of interest and a short list of not more than three tenderers is recommended. Moreover, it is commonplace in the large civil construction industry to compensate tenderers for the reasonable cost of preparing a tender design. Consequently, it is recommended that an allowance for the cost of mixture design preparation be offered to short-listed tenderers.

4.4 Contractual Provisions

The recommended approach to delivering runway resurfacing works under a design and construct contract requires a number of particular contractual provisions. These include:

- Design responsibility vested in the contractor. Primarily for the contractor to convert the preliminary design to a final design, to be accepted by the airport.
- Pricing reconciliation mechanism. To reflect the reliance on the nominated quantities during the tender phase and the deviations implied by the final design.
- Warranty schedule. To reflect the contractor's responsibility for the field performance of the asphalt surface extending beyond the standard defects liability period. The basis of the warranty is detailed further below.

The warranty schedule is required to contractually formalise and define the contractor's liability for surface performance. The duration of the warranty must be appropriate to extend beyond the timeframe during which significant distress is likely to be observed. However, the timeframe until significant maintenance is expected, such as asphalt preservation is best avoided. Reports of runway asphalt distress indicate that symptoms generally appear within two to three years of resurfacing. The critical distresses, such as shoving and groove closure, most often present during the first hot summer weather after resurfacing. Therefore, a five-year warranty period, from the completion of the asphalt resurfacing works, is recommended, which remains well short of the typical seven to eight-year timeframe until asphalt preservation is considered.

It is important to understand that the warranty applies to the performance of the surfacing and not to the pavement as a whole. If the pavement has also been reconstructed or rehabilitated under the same contract, standard design liabilities and the normal defects liability period apply to the non-surface portion of the pavement. Consequently, the warranty should exclude liability for distresses caused by the underlying pavement layers or gross structural overload by larger than expected aircraft. Although difficult to pre-define these excluded distresses, issues are generally readily attributable to either the surface of the underlying pavement by post-distress forensic investigation. It is also important to acknowledge that the warranty period is not intended to be a maintenance-free period. Rather, the warranty is intended to be triggered by excessive surface distresses that significantly reduce the service life of the surface. Routine reactive maintenance is essential to ensuring that a normal and minor distress, such as a crack, does not develop into a major defect.

4.5 Maintenance Inputs

Contractors rightly expect airports to appropriately maintain runway surfaces during the warranty period. It is important that the contractor be involved in the surface condition assessment and subsequent maintenance planning. It is also preferable that the contractor delivers the maintenance, although this is not always practicable or cost-effective.

The most important maintenance activity is sealing cracks, performed without delay, to prevent moisture from entering the asphalt surface and underlying pavement. It is recommended that the contractor be paid an annual sum to attend a condition inspection, prepare maintenance recommendations and execute unlimited crack sealing in a single annual visit. Other periodic maintenance activities include linemarking and rubber tyre rubber contamination removal. These activities are often undertaken by local or specialised service providers rather than the asphalt surface constructor. It is important to protect the warranty provisions by ensuring that appropriate linemarking paint is selected and that rubber removal is performed to minimise the impact on the asphalt surface layer.

5 Summary and Conclusions

The traditional approach to runway resurfacing used a prescriptive and compliance-based asphalt specification in a construct-only contractual framework. Challenges associated with compliant runway surfaces failing to perform as expected has led to the development of a performance-based asphalt specification for Australian airports. Although this specification can be used for the asphalt element of a construct only package of design documentation, a design and construct approach reduces the risk of asphalt complying with a sound design not performing in the field and becoming the airport's responsibility to fix. The design and construct approach is expected to become more common in the future, but a preliminary design and appropriate tendering/contractual conditions are critical to realising the potential benefits.

Acknowledgements. The support of Dubbo Airport (New South Wales) and Whitsunday Coast Airport (Queensland) in the development of the performance based asphalt specification and design and construct runway resurfacing framework is gratefully acknowledged and greatly appreciated.

References

Diefenderfer, B.K., Bryant, J.W.: Development of a pavement warranty contract and performance specification for hot-mix asphalt resurfacing project. In: 2006 Airfield and Highway Pavements Specialty Conference, Atlanta, Georgia, USA, 1–3 May 2006

Emery, S., Man, I., Mihaljevic, I.: New modified binder for airport asphalt. In: Proceedings AAPA International Flexible Pavements Conference, Gold Coast, Queensland Australia. Australian Asphalt Pavement Association, 13–16 Sept 2015

Jansz, R.: An airport surfacing binder and mix for Australia. In: 17th AAPA International Flexible Pavements Conference, Melbourne, Victoria, Australia, 13–16 Aug 2017

Rodway, B.: Flexible aircraft pavement surfacing—Australian practice. In: 8th International Conference on Maintenance and Rehabilitation of Pavements, Singapore, 27–29 July 2016

White, G.: Warm mix asphalt for Australian airports. Int. J. Pavement Eng. Asphalt Technol. **16** (1), 11–29 (2015)

White, G.: Challenges for Australian flexible airport pavements. Aust. Geomech. **51**(3), 39–46 (2016a)

White, G., Thompson, M.: Australian airport asphalt surface treatments. In: 6th Eurasphalt & Eurobitume Congress, Prague, Czech Republic, 1–3 June 2016

White, G.: Modification of the airport pavement strength rating system for improved protection of asphalt surfaces. Int. J. Pavement Eng. (2017a). https://doi.org/10.1080/10298436.2017.1316641. Article-in-press

White, G.: State of the art: asphalt for airport pavement surfacing. Int. J. Pavement Res. Technol. (2017b). https://doi.org/10.1016/j.ijprt.2017.07.008. Article-in-press

White, G., Embleton, K.: Next generation binder for airport asphalt. In: AAPA International Flexible Pavements Conference, Gold Coast, Queensland, Australia. Australian Asphalt Pavement Association, 13–16 Sept 2015

White, G.: Changes in Australian paving-grade bitumen: are they real and what should Australia do about it? In: 27th ARRB Conference, Melbourne, Victoria, Australia, 16–18 Nov 2016b

White, G.: Development of a performance specification for airport asphalt. In: 17th AAPA International Flexible Pavements Conference, Melbourne, Victoria, Australia, 13–16 Aug 2017d

White, G.: Expedient runway upgrade technologies. In: 10th International Conference on the Bearing Capacity of Roads, Railways and Airfields, Athens, Greece, 28–30 June 2017e

White, G.: Towards a performance-based airport asphalt specification. In: International Conference on Highway Pavements and Airfield Technology, Philadelphia, Pennsylvania, USA, 27–30 Aug 2017c

Influence of FWD Uncertainty on Back-Calculated Flexible Pavement Layer Moduli

José Neves[1(✉)] and Edgar Cardoso[2]

[1] CERIS, Department of Civil Engineering, Architecture and Georesources, Instituto Superior Técnico, Universidade de Lisboa, Lisbon, Portugal
jose.manuel.neves@tecnico.ulisboa.pt
[2] Instituto Superior Técnico, Universidade de Lisboa, Lisbon, Portugal

Abstract. The Falling Weight Deflectometer (FWD) is one of the most important non-destructive tests (NDT) used for pavement structural evaluation based on deflection measurements. Inverse analysis of FWD measurements—back-calculation—is a common process for obtaining the pavement layer moduli, including the subgrade soil, which are useful for estimating remaining pavement life for overlay design purposes. Based on a proficiency test (PT) involving testing devices from different manufacturers, the objective of this study is to analyze the influence of FWD uncertainty on the back-calculated flexible pavement and foundation layers moduli. A parametric study was performed considering the back-analysis of different asphalt pavements with regard to layer thickness and resilient modulus (unbound granular and asphalt concrete layers, including subgrade). It was concluded, in general, that FWD uncertainty could have an important influence on the layers modulus obtained from back-calculations. This influence was less significant in the case of asphalt concrete layers (±25%). The variation in resilient modulus was greater in the layers of the subgrade soil (−33 to +83%) and the unbound granular materials (−42 to +80%).

1 Introduction

The Falling Weight Deflectometer (FWD) is the most widely used non-destructive test (NDT) device for assessing pavement bearing capacity (Bush and Baladi 1989; Quintus et al. 1994; Tayabji and Lukanen 2000). FWD consists of applying a dynamic load to the pavement, generated by a weight dropping along a guide system on a buffer, which is transmitted through a circular plate on the pavement surface. The resulting deflections on the pavement surface are measured using sensors mounted on the surface of the pavement. A deflection basin is obtained from the impact load.

The back-calculation process consists in obtaining an estimated deflection basin, by means of numerical analysis, that approaches the measured deflection basin. Irwin (2002) conducted a thorough review of some back-calculation models that are currently and commonly used to evaluate the properties of the pavement structure from the measured surface deflection basin. However, the back-calculation process is mathematically a complex inverse method that can be applied to the estimation of layers

Z. Hossain et al. (eds.), *Solving Pavement and Construction Materials Problems with Innovative and Cutting-edge Technologies*, Sustainable Civil Infrastructures,
https://doi.org/10.1007/978-3-319-95792-0_13

moduli. At least three major sources of uncertainty can contribute to inaccuracies in the back-calculated moduli based on FWD: (1) uncertainty concerning the measured parameters (deflections); (2) simplifications used in the back-calculation method; and (3) random deviations of real pavement parameters (layer thickness and mechanical properties) from those assumed or specified (Ahmed 2006).

Considering the influence of the uncertainty of deflection measurements on back-calculation accuracy, several authors have studied FWD limitations, mainly out of concern as to the lack of reproducibility between individual equipment (Irwin 2002; Chen et al. 1999; Lukanen 1992; Murphy 1998; Van Gurp 1991). Rocha et al. (2004) presented a review on the accuracy and precision of FWD. Various FWD are commercially available and, according to a number of authors, each device exhibits a reasonable degree of repeatability but, in general, a lower reproducibility exists between different devices. The sources of non-reproducibility derive from multiple factors involved in the testing process that are sometimes impossible to control and mitigate. The load pulse shape is dependent on the buffer system geometry and alignment, but also influenced by the pavement and subgrade constituting material. The load pulse duration is currently sometimes determined by factory default. The manufacturer's technique used for signal conversion also influences the quality of the measured deflection data. High frequency disturbance can contribute to distortion of the pulse load shape and the deflection pulse. Comparing distorted peak values of the load time history and deflection time histories may affect back-calculation results. It is therefore recommended, when available, to enable the smoothing filter function to cut off frequencies above 60 Hz (Sorensen 1993; Van Gurp 1995).

Simplifications of the back-calculation method and deviations of real pavement structures are the focus of the literature on the accuracy of layers moduli evaluation of the existing pavements. However, there is a gap in the literature with regard to values for the precision of the FWD deflections and the resulting uncertainty. The objectives of this paper are: (1) to present values for FWD precision and uncertainty; (2) to evaluate the influence of the deflection uncertainty on the accuracy of back-calculated pavement layers moduli.

The deflection uncertainty was based on the repeatability and reproducibility obtained in a proficiency test (PT) that consisted in an interlaboratory comparison involving laboratories with testing devices from different manufacturers. The paper presents a sensitivity analysis of back-calculation result accuracy in terms of FWD uncertainty. A parametric study was performed considering the back-analysis of different flexible pavements regarding layer thickness and resilient modulus (unbound granular and asphalt concrete layers).

2 Methodology

2.1 Deflection Uncertainty

Uncertainty of deflection measurements was based on the repeatability and reproducibility obtained in a PT organized with three different FWD devices: Carl Bro PRI 2100 FWD trailer; KUAB 240 HWD trailer; and Dynatest 8002 FWD trailer. The testing site was a flexible pavement composed of asphalt concrete AC14 in the surface layer (5 cm) and unbound granular material UGM in the subbase layer (20 cm). The sequence of tests was defined with a view to minimizing variations in the asphalt concrete temperature. The PT was organized in line with the requirements of ISO/IEC 17043 (Neves and Cardoso 2017).

FWD tests have followed the method described in ASTM D4694 (2015) and ASTM D4695 (2015). Five loading sequences were always performed and deflection average of the last three sequences was calculated. The load peak was controlled in each device in order to obtain 65 and 90 kN through a circular plate with a diameter of 300 mm. The deflections were measured by the sensors at 8 points located at different distances from the load: 0, 30, 45, 60, 90, 120, 150 and 180 cm. Figure 1 presents an example of the deflection basins measured during the tests. The standard deviation of measured deflections is represented for each sensor point. Taking into account the vertical scale, the standard deviations are represented with an amplification of ten times because the values were, generally speaking, very small, which is indicative of good repeatability. In general, one can conclude that there was a tendency for one of the devices—FWD 2—to obtain lower deflection values. This particular behavior of FWD 2 had a significant influence on the reproducibility quality.

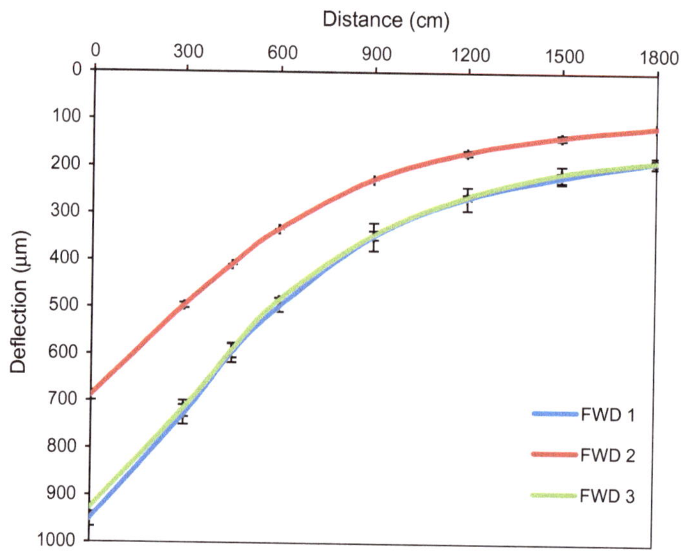

Fig. 1. Example of deflection basins measured for 90 kN

Repeatability was analyzed on the basis of the last three sequences of load tests: using the same procedure, device and test location. The reproducibility was assessed by mean of the use of the three different FWD devices. The repeatability and reproducibility of the deflections were analyzed in accordance with ISO 5725-2 (1994). Prior to precision analysis, deflections were normalized to the appropriate nominal load levels, i.e. 65 and 90 kN. Figure 2 represents the calculated limits of repeatability (r) and reproducibility (R) (confidence level of 95%), as a function of deflection magnitude (D) (Fig. 2b). The results confirm, in general, a reasonable level of repeatability (Fig. 2a) but poor reproducibility (Fig. 2b), mainly due to one of the testing devices (FWD 2), as previously pointed out (Neves and Cardoso 2017). From Fig. 2, Eqs. (1) and (2) for the limits of precision were obtained by the regression of the calculated values.

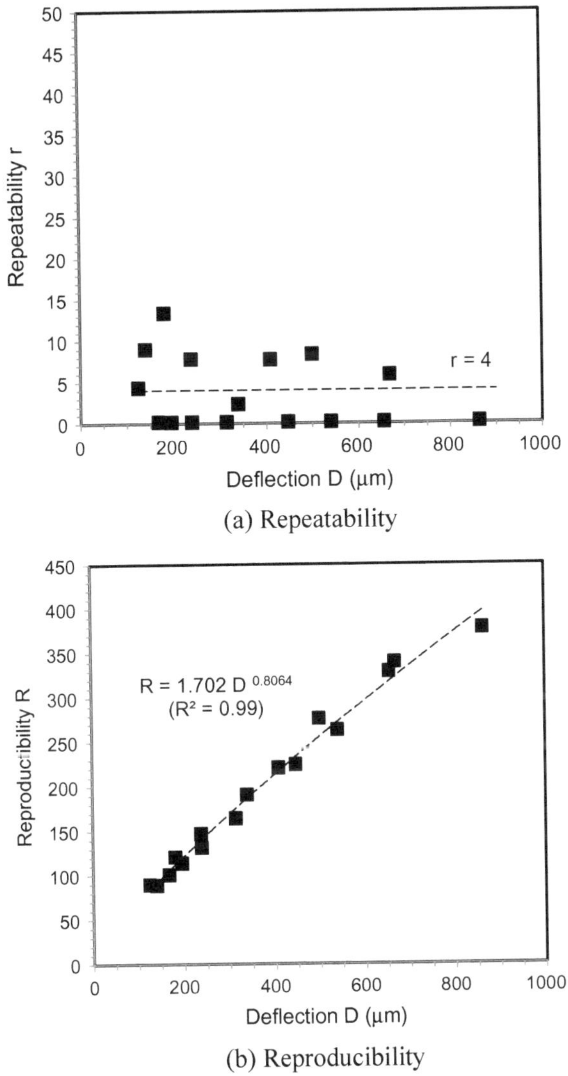

Fig. 2. Precision of deflections

$$r = 4 \tag{1}$$

$$R = 1.702D^{0.8064} \tag{2}$$

Uncertainty of a measurement is the dispersion of the quantity values that could be attributed to the measured parameter. The estimation of the uncertainty was based on the standard deviations of the repeatability and reproducibility. Figure 3 presents the uncertainty (U) as a function of the deflection magnitude (D) for a confidence level of 95%. The regression of the results resulted in Eq. (3) which can be useful to calculate the critical range of values (D_{lim}) for a certain deflection magnitude (D), as indicated by Eq. (4).

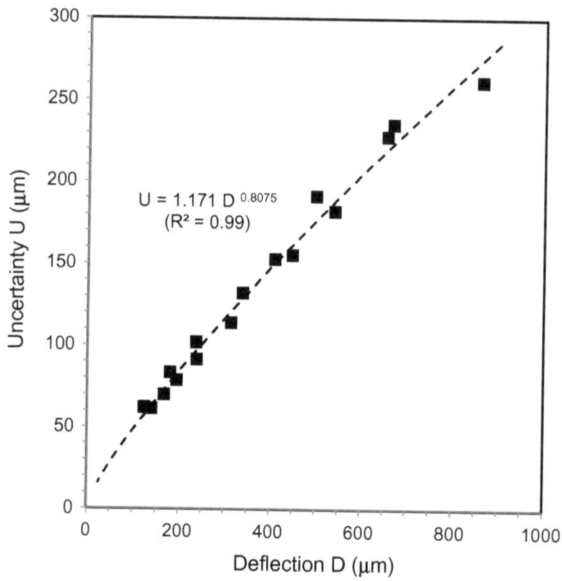

Fig. 3. Uncertainty of deflection measurements

$$U = 1.171D^{0.8075} (\mu m) \tag{3}$$

$$D_{lim} = D \pm U \tag{4}$$

2.2 Sensitivity Analysis

The study of the influence of deflection uncertainty on the back-calculated flexible pavement and foundation layers moduli was based on a sensitivity analysis that took into consideration several flexible pavement structures selected from the Portuguese catalogue (Neves and Cardoso 2017). Table 1 shows the main properties of the

Table 1. Properties of pavements

Layer materials		Properties	Values			
AC		t_{AC} (cm)	16–32	18–28	12–28	10–26
		E_{AC} (MPa)	4000		4000	
UGM	Base	t_{UGM} (cm)	n.e.		20	
		E_{UGM} (MPa)	n.e.		400	
	Subbase	t_{UGM} (cm)	20		20	
		E_{UGM} (MPa)	200		200	
Subgrade soil		E_S (MPa)	60	100	60	100

n.e.—not existent; AC—asphalt concrete; UGM—unbound granular materials; *t*—thickness; *E*—resilient modulus

pavement structures: layers and materials (asphalt concrete (AC) in upper layers, unbound granular materials (UGM) in base and subbase layers); subgrade soil; thickness of asphalt concrete (t_{AC}) and unbound granular materials (t_{UGM}); resilient modulus of asphalt concrete (E_{AC}), unbound granular materials (E_{UGM}) and subgrade soil (E_S). An elastic linear analysis was performed in a back-analysis process considering the mechanical properties of the materials used in pavement layers: resilient modulus and Poisson ratios. Poisson ratios adopted were 0.35 for asphalt concrete and 0.30 for unbound granular material and subgrade soil.

Figure 4 shows an example of the methodology used to generate the deflection basin for each pavement structure and, consequently, the deflection limits with regard to the corresponding uncertainty resulting from Eqs. (1) and (2). Figure 4a presents the geometrical and mechanical properties of one pavement used in the sensitivity analysis (Table 1). Figure 4b shows the reference deflection basin obtained when a vertical force of 65 kN was applied on a rigid and circular plate of 15 cm radius resting on the pavement surface. The vertical lines associated to each deflection represent the uncertainty calculated by Eq. (1). These limits of deflection basins were to be used in the back-analysis methodology.

The back-analysis procedure consisted in evaluating the mechanical properties of the pavement structures from the limits of the deflection basin caused by the static loading. A linear elastic multilayered model with a constant set of parameters was assumed, the parameters being: a static load, linear elastic modulus and Poisson's ratios of each layer, and layer thickness. It was also assumed that the layers were homogeneous and evenly thick throughout their length. The BISAR program was used for the calculations. An iterative and manual step sequence was used to refine these parameters until the deflection basin closely matched the reference deflection basin. Quality of deflections adjustment was based on the root mean square error (RMSE), which was kept under 4% throughout.

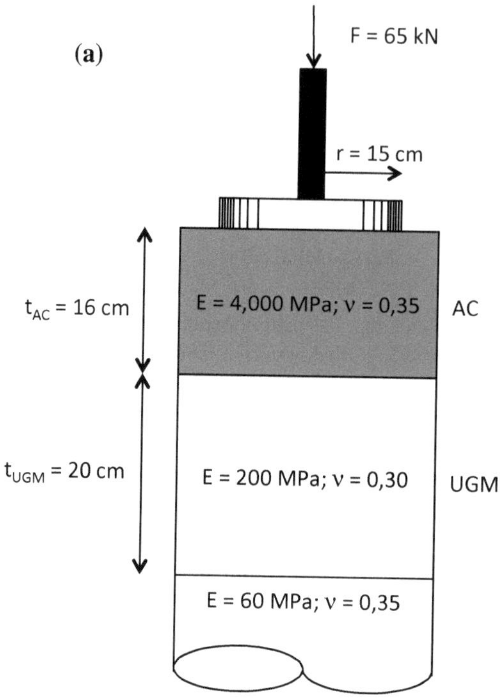

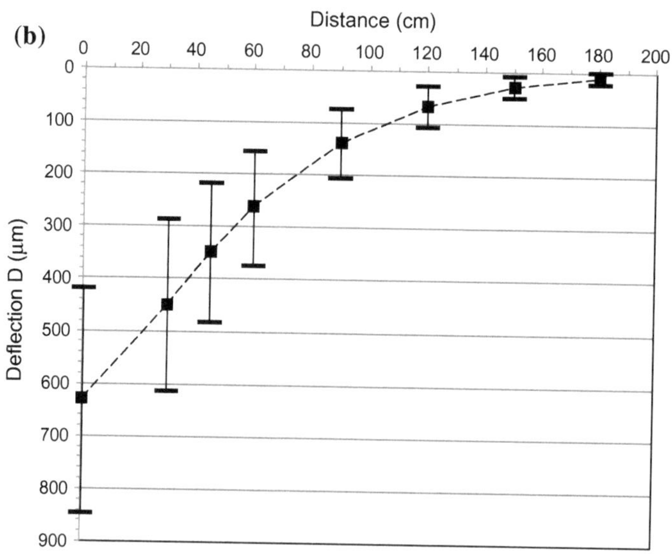

Fig. 4. Example of data used in back-analysis methodology

3 Results and Discussion

The results with regard to FWD precision—repeatability and reproducibility—are in line with the reviewed literature (Garg 2002; Murphy 1998; Rocha et al. 2004). Although repeatability is normally achieved, reproducibility is dependent on ambient factors and the features of each FWD, such as the deflectometer type, the analog-to-digital signal interpretation algorithm proprietary of each manufacturer, the quality of the rubber underneath the loading plate, the shape and quality of the rubber buffers, the load pulse history, the drop height, the force frequency, and many other variables that make reproducibility hard to control effectively. Neves and Cardoso (2017) have presented experiment-based research concerning the FWD precision. This study confirmed a satisfactory repeatability of deflection measurements. In contrast, reproducibility was difficult to achieve in most cases. Consequently, the uncertainty was revealed to be at a high level. Uncertainty and precision revealed to be dependent on deflection magnitude: uncertainty increased the higher the deflections.

The back-calculation analysis performed under FWD uncertainty resulted in a range of interval for estimated moduli, representing the type of varying results that may be obtained with regard to reproducibility issues reported by the literature. The parametric analysis produced a study of the deflections measured in different pavement structures for the two peak nominal loads (65 and 90 kN) that were back-calculated to obtain layers moduli. The result was a converging set of moduli intervals for each pavement layer.

Figure 5 represents the variation in percentage (ΔE) of the back-calculated moduli of pavement layers (E_{BC}) in relation to the reference values presented in Table 1 (E) (Eq. 5). The results are represented as a function of the thickness ratio (t_{AC}/t_{TOTAL}) between the total thickness of asphalt concrete layers (t_{AC}) and the total pavement thickness (t_{TOTAL}).

$$\mathbf{\Delta E} = \left(\frac{\mathbf{E_{BC}}}{\mathbf{E}} - \mathbf{1}\right) \times \mathbf{100}(\%) \tag{5}$$

Analysis of Fig. 5 allows for the following main conclusions:

- In general it was observed that FWD uncertainty had an important effect on the pavement layers moduli, including subgrade soil, obtained from back-calculations;
- The influence was less significant in the case of asphalt concrete layers (±25%) (Fig. 5a);
- Variation in resilient modulus was greater in the layers of unbound granular materials (−42 to +80%) (Fig. 5b) and the subgrade soil (−33 to +83%) (Fig. 5c);
- Regarding the influence of layer thickness, a certain slight tendency was observed towards greater sensitivity to FWD uncertainty for thicker asphalt layers. This effect was more significant in the case of a positive variation in the back-calculated resilient modulus of unbound layers of base, subbase and subgrade (Fig. 5b, c).
- Taking into account that back-calculation is a common process used to estimate pavement remaining life in the case of overlay design, a positive variation on the layers moduli can result in overestimated structural behavior of the existing pavement. This effect can have a negative impact on the pavement reinforcement.

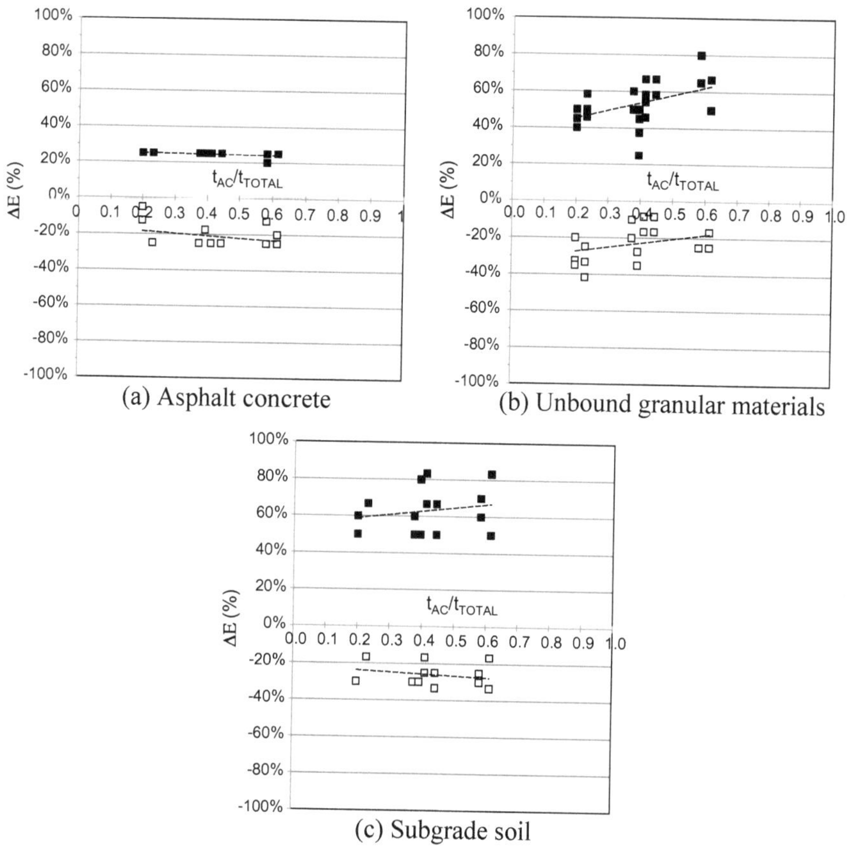

Fig. 5. Effect of deflection uncertainty on back-calculated layers moduli

4 Conclusions

The FWD is a common NDT device used for pavement structural evaluation. This test is essential to achieve a mechanistic approach to the existing pavement structure for rehabilitation by back-calculation analysis of test results. The paper has analyzed the influence of FWD test uncertainty, evaluated on the basis of a PTS organized in accordance with ISO/IEC 17043 (2010), on the back-calculated pavement and foundation layers moduli. Assessment of uncertainty was based on the repeatability and reproducibility obtained with different FWD devices on a flexible pavement and following the procedures of ASTM D4694. The estimation of the deflections uncertainty was based on the standard deviations of the repeatability and reproducibility obtained in accordance with the methodology of ISO 5725-2. Due to poor reproducibility, the uncertainty of deflections could be important depending on the deflection magnitude.

A sensitivity analysis was performed by means of a parametric study considering different structures of flexible pavements (layer thickness and stiffness). A range of deflections was calculated for these pavements. The back-analysis of those deflections made it possible to obtain the variation of pavement layers moduli (asphalt concrete and unbound granular layers), including the subgrade soil. It was concluded in general that FWD uncertainty could have a significant influence on the layers moduli obtained from back-calculations. This influence was less significant in the case of asphalt concrete layers (±25%). Variation in resilient modulus was greater in the layers of the subgrade soil (−33 to +83%) and the unbound granular materials (−42 to +80%). A higher sensitivity to FWD uncertainty was observed in the case of thicker asphalt layers and for a positive variation in the back-calculated resilient modulus of the base, subbase and subgrade layers. Given that back-calculation is a process commonly used to estimate pavement remaining life, FWD uncertainty can, as indeed expected, have an important impact on the pavement overlay design.

Acknowledgements. The authors wish to thank RELACRE, the Portuguese Association of Accredited Laboratories, for the permission to use data from FWD proficiency tests.

References

Ahmed, N.G.: Influence of variability in flexible pavement parameters on backcalculated moduli. J. Eng. **12**(2), 445–457 (2006)

ASTM D 4694-09: Standard test method for deflections with a falling-weight-type impulse load Device. American Society for Testing and Materials, USA (2015)

ASTM D 4695-03: Standard guide for general pavement deflection measurements. American Society for Testing and Materials, USA (2015)

Bush, J.A., Baladi, G.Y.: Nondestructive testing of pavements and backcalculation of moduli. In: International Symposium on Nondestructive Testing of Pavements and Backcalculation of Moduli, Baltimore, Maryland, USA (1989)

Chen et al.: Effects of buffers on falling weight deflectometer Measurements. Internal Technical Memo, Design Pavement Section, TxDOT, Austin, Texas, USA (1999)

Garg, N.: Comparison between falling weight deflectometer and static deflection measurements on flexible pavements at the national airport pavement test facility (NAPTF). Federal Aviation Administration (2002)

Irwin, L.H.: Backcalculation : an overview and perspective. FWD backcalculation. In: Workshop 3, 6th International Conference on the Bearing Capacity of Roads, Railways and Airfields, Cascais, Portugal (2002)

ISO 5725-2: Accuracy (trueness and precision) of measurement methods and results. Part 2: Basic method for the determination of repeatability and reproducibility of a standard measurement method. International Organization for Standardization (1994)

ISO/IEC 17043: Conformity assessment—general requirements for proficiency testing. International Organization for Standardization (2010)

Lukanen, E.O.: Effects of buffers on falling weight deflectometer loadings and deflections. Transp. Res. Rec. (1355), 37–51 (1992)

Murphy, M.: A mechanistic-empirical approach to characterizing subgrade support and pavement structural condition for network-level application. Ph.D. dissertation, University of Texas, Austin, USA (1998)

Neves, J., Cardoso, E.: Uncertainty evaluation of deflection measurements from FWD tests on road pavements. In: 7th International Conference on Engineering Surveying, Lisbon, Portugal (2017)

Quintus, et al.: Nondestructive testing of pavements and backcalculation of moduli. ASTM special technical publications, Phladelphia, PA. USA (1994)

Rocha, et al.: Falling weight deflectometer fleet repeatability and reproducibility. Road Mater. Pavement Des. **5**(2), 215–238 (2004). https://doi.org/10.1080/14680629.2004.9689970

Sorensen, A.: Dynatest FWD/HWD field programs—smoothed (signal) peak option. In: 2nd FWD Seminar of Forum of European National Highway Research Laboratories (FEHRL), Laboratoire Central des Points et Chaussées, France (1993)

Tayabji, S., Lukanen, E.O.: Nondestructive Testing of Pavements and Backcalculation of Moduli, vol. 3. ASTM special technical publication, West Conshohocken, PA (2000)

Van Gurp, C.: Characterization of seasonal influences on asphalt pavements with the use of falling weight deflectometers. PhD dissertation, Technische Universiteit Delft (1995)

Van Gurp, C.: Consistency and reproducibility of falling weight deflections. In: Conference on Road and Airport Pavement Response Monitoring Systems, West Lebanon, New Hampshire, USA, pp. 291–305 (1991)

Feasibility of Using XRF for Assessment of Surface Free Energy Components of Asphalt Binder

Syed Ashik Ali[1(✉)], Rouzbeh Ghabchi[2], Shivani Rani[1], Mohammed Ashiqur Rahman[1], Musharraf Zaman[3], and Edgar A. O'Rear[4]

[1] School of Civil Engineering and Environmental Science, The University of Oklahoma, 202 West Boyd Street, Room 334, Norman, OK 73019, USA
{syed.a.ali, shivani.rani}@ou.edu, shovon0701001@gmail.com

[2] Department of Civil and Environmental Engineering, South Dakota State University, 132 Crothers Engineering Hall, Brookings, SD 57007, USA
rouzbeh.ghabchi@sdstate.edu

[3] Southern Plains Transportation Center, the University of Oklahoma, 202 West Boyd Street, Room 334, Norman, OK 73019, USA
zaman@ou.edu

[4] School of Chemical Biological, and Materials Engineering, Institute for Biomedical Engineering, Science and Technology, University of Oklahoma, Norman, OK 73019, USA
orear@ou.edu

Abstract. Recently the surface free energy (SFE) techniques are being used to mechanistically quantify bonding characteristics of aggregate-asphalt binder systems and consequently moisture-induced damage potential of asphalt mixes. Also, the chemical composition of the asphalt binder is known to be an important factor in determining the strength of the asphalt-aggregate bond. The present study was undertaken to explore correlations between the chemical compositions of the binders containing different additives with their SFE components. For this purpose, a PG 64-22 asphalt binder was collected from a local supplier in Oklahoma. A commonly used chemical warm-mix asphalt (WMA) additive and an anti-stripping agent (ASA) were also collected and blended with the PG 64-22 binder at two different proportions, namely 0 and 0.5% by the weight of the asphalt binder. The SFE components of the short-term aged asphalt binders blended with additives were determined using Dynamic Wilhelmy Plate (DWP) test. The elemental analysis of the binders was carried out using an X-ray Fluorescence (XRF) analyzer. From the DWP tests it was concluded that the additions of WMA additive and ASA had significant effects on the values of the SFE components of the neat binder. The XRF results are expected to help explain the changes in SFE components due to any additive in an asphalt binder. Also, this study is expected to help predict the surface free energies of the binder from its chemical composition using a less time-consuming XRF test.

Z. Hossain et al. (eds.), *Solving Pavement and Construction Materials Problems with Innovative and Cutting-edge Technologies*, Sustainable Civil Infrastructures,
https://doi.org/10.1007/978-3-319-95792-0_14

1 Introduction

Moisture-induced damage is one of the major distresses which contributes to the significant premature deterioration of asphalt pavements. The loss of strength and durability in asphalt mixes due to the reduction in bond strength between aggregate and binder in presence of moisture is called moisture-induced damage (Masad et al. 2006; Bhasin et al. 2007; Lu and Hervey 2008). Many State Departments of Transportation are spending significant amounts of resources to combat Moisture-induced damage problems. Generally, modified Lottman test, indirect tensile strength ratio (TSR), resilient modulus ratio, Marshall stability ratio, stripping inflection point (SIP) from Hamburg wheel tracking (HWT) test and fracture energy ratio are used for evaluating the moisture-induced damage potential of asphalt mixes (Bagampadde et al. 2006; Gorkem and Sengoz 2009; Ghabchi et al. 2015; Mirzababaei 2016). However, none of these test methods addresses the failure mechanisms governing the moisture-induced damage of asphalt pavements. From a mechanistic point of view, the moisture-induced damage can be assessed by evaluating the bond strength between asphalt binder and aggregate, in presence of moisture. A better resistance to moisture-induced damage potential can be ensured by improving the adhesion bonding in asphalt-aggregate system (Masad et al. 2006; Lu and Hervey 2008). Hefer et al. (2005) and Caro et al. (2008) mentioned the following five different theories to explain the adhesion bonding in asphalt-aggregate system: weak boundary layer theory, electrostatic theory, chemical bonding theory, mechanical bonding theory and thermodynamic theory. A number of recent studies has used the thermodynamic theory or adhesion due to surface free energy (SFE) approach to mechanistically quantify the bonding between aggregate and binder and hence evaluated the moisture-induced damage potential of asphalt mixes (Bhasin et al. 2006, 2007; Hefer et al. 2006; Wasiuddin et al. 2007, 2008; Buddhala et al. 2011; Ghabchi et al. 2013). Also, the SFE technique was used successfully to evaluate the changes in the moisture-induced damage potential of asphalt mixes containing different additives (Wasiuddin et al. 2007; Moghadas et al. 2012; Arabani and Hamedi 2014).

The chemical characteristics of a binder are known to be an important factor in determining the strength of the asphalt-aggregate bond. Several techniques such as X-Ray fluorescence (XRF), Fourier transform infrared (FTIR) spectroscopy, differential scanning calorimetry, nuclear magnetic resonance (NMR), and X-ray photo electron spectroscopy (XPS) are being used to determine the chemical constituents of asphalt binder and aggregate (Le Guern et al. 2010; Hossain et al. 2012; Hesp and Shurvell 2013). These techniques were found to provide important information about the chemical properties of asphalt binder with the aggregate within a short amount of time. Several studies used the XRF technique for the quality assurance of asphalt binder modified with waste engine oil (Soleimani et al. 2009; Hesp and Shurvell 2010, 2013). In the present study, elemental analyses on selected binders were carried out using an XRF in order to correlate the chemical composition of the binders with their bonding properties.

Recently the use of warm mix asphalt (WMA) for construction of pavements has been increased to reduce energy consumption, preserve the environment and ensure sustainable development. The WMA technologies are reported to reduce the asphalt production temperature by 2–38 °C than that of hot mix asphalt (HMA) which results in a significant savings of fuel costs (D'Angelo et al. 2008; West et al. 2014). Three categories of WMA technologies, namely, asphalt foaming, organic additives and chemical additives are currently available to the asphalt industry (West et al. 2014).

The purpose of the present study is to investigate the effects of a chemical WMA additive and an antistripping additive on the surface characteristics of a binder. The specific objectives of this study are to:

i. Determine the effects of a chemical WMA additive on the SFE components of a binder;
ii. Determine the SFE components of a binder blended with an antistripping agent;
iii. Determine the chemical components of asphalt binder blends using XRF technique; and
iv. Determine correlations between the SFE components and the chemical compositions of binders to estimate the moisture-induced damage potential of asphalt binder.

2 Surface Free Energy (SFE)

The surface free energy (SFE) can be defined as the amount of work required to increase the surface of a solid by a unit area under vacuum (Van Oss et al. 1988). The SFE of a material comprises of three components, namely an apolar or Lifshitz-van der Waals component, a monopolar acidic component, and a monopolar basic component (Van Oss et al. 1988). The total SFE of a material can be expressed by Eqs. (1) and (2).

$$\Gamma^{Total} = \Gamma^{AB} + \Gamma^{LW} \quad (1)$$

$$\Gamma^{AB} = 2 \times \sqrt{(\Gamma^{+}\,\Gamma^{-})} \quad (2)$$

where,

Γ^{Total} = total SFE of the material,
Γ^{AB} = acid-base component,
Γ^{LW} = Lifshitz-van der Waals component,
Γ^{+} = Lewis acid component, and
Γ^{-} = Lewis base component.

The SFE components of an asphalt binder can be determined by measuring the contact angle with three different solvents namely, one apolar, one monopolar and one bipolar (Bhasin et al. 2007) of known SFE components. The solutions of the system of three simultaneous equations in the form of Eq. (3) will resulted in SFE components of asphalt binder.

$$\Gamma_{L}(1 + \cos\theta) = 2 \times \sqrt{(\Gamma_{A}^{LW}\,\Gamma_{L}^{LW})} + 2 \times \sqrt{(\Gamma_{A}^{+}\,\Gamma_{L}^{+})} + 2 \times \sqrt{(\Gamma_{A}^{-}\,\Gamma_{L}^{-})} \quad (3)$$

where, θ is the contact angle between asphalt binder and solvent. The Γ_{L}^{LW}, Γ_{L}^{+} and Γ_{L}^{-} represent the Lifshitz-van der Waals component, acid component and base component of the liquid solvent.

3 Materials and Methods

3.1 Materials

For the purpose of this study, a PG 64-22 asphalt binder was collected from a local refinery in Oklahoma. Also, a commonly used and commercially available chemical WMA additive (W1) and an antistripping additive (A1) were collected from its vendor. The additives were added to the asphalt binder at two different amounts, namely 0 and 0.5% by weight of asphalt binder. The mixing of the neat binder and the additives were conducted using a high shear mixer at a speed of 1000 rpm for 45 min at a temperature of 155 °C. The Rolling Thin Film Oven (RTFO) aging was conducted according to the AASHTO T 240-13 test method to simulate the oxidation and aging of binders during mixing in the asphalt plant and compacting in the field. For convenience, the PG 64-22 binder with 0.5% WMA additive and 0.5% antistripping additive is represented as PG 64-22 + 0.5% W1 and PG 64-22 + 0.5% A1, respectively.

3.2 Methods

3.2.1 Dynamic Wilhelmy Plate (DWP) Test

The Dynamic Wilhelmy Plate (DWP) test was used here to determine the SFE components of the PG 64-22, PG 64-22 + 0.5% W1 and PG 64-22 + 0.5% A1 binders. The contact angles of the asphalt binder blends with three different probe liquids, namely water, glycerin and formamide were measured using a dynamic contact angle analyzer (DCA). The details of the testing procedure are described elsewhere (Ghabchi et al. 2013). At least five samples were tested for each solvent to ensure consistency and repeatability of the test results. In this study, a total of 45 samples were tested for contact angle measurements. The advancing contact angles were measured for determining the SFE components of asphalt binder blends as they were found to be more consistent than the measured receding contact angles (Hefer et al. 2006). Figure 1a, b present the photographic views of the asphalt binder samples and DWP test in progress, respectively.

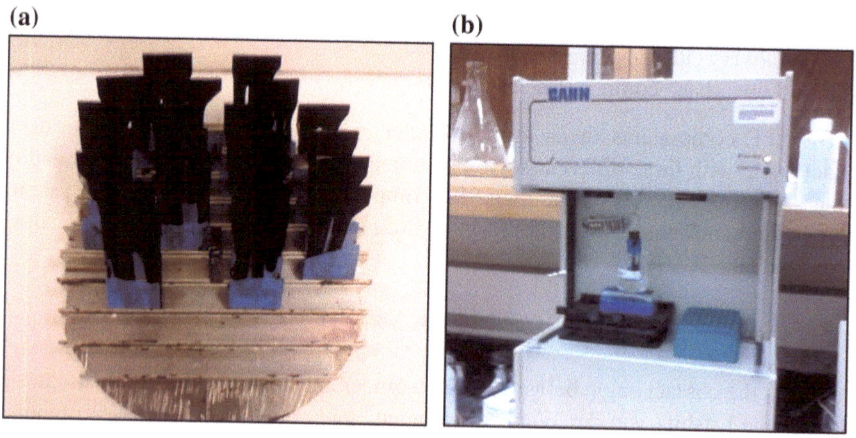

Fig. 1. **a** Binder samples for DWP test; and **b** DWP test in progress

3.2.2 XRF Test on Asphalt Binders

The elemental analysis of PG 64-22, PG 64-22 + 0.5% W1 and PG 64-22 + 0.5% A1 binders were conducted using X-ray fluorescence (XRF) technique. In XRF, the emission of characteristic "secondary" (or fluorescent) X-rays from the test sample excited by high-energy X-rays is used to identify the chemical compositions of materials being studied. The working principles and mechanism of the XRF technique can be found elsewhere (Hesp and Shurvell 2010, 2013). All the tests were conducted in the chemical laboratory of Ingevity using a Rigaku NexCG X-Ray Fluorescence Device. The device is capable of providing rapid, non-destructive, multi-element analyses from very low to high concentrations of elements ranging from sodium (Na) to uranium (U). Peak heights in the spectrum were used to detect and quantify the presence of the elements. Three samples were tested from each binder blends to ensure consistency. Table 1 presents the test matrix for asphalt binder blends of this study.

Table 1. Test matrix of the binder blends

SFE of asphalt binders		
Material	Solvent type	No. of samples
PG 64-22	Water	15
PG 64-22 + 0.5% A1	Glycerin	15
PG 64-22 + 0.5% W1	Formamide	15
XRF testing of binders		
PG 64-22		3
PG 64-22 + 0.5% A1		3
PG 64-22 + 0.5% W1		3

4 Results and Discussions

4.1 Surface Free Energy of Asphalt Binders

4.1.1 Contact Angles of Asphalt Binders

In this study, the contact angles of the asphalt binder blends with three different probe liquids were used to calculate the SFE components of the corresponding binders. The contact angles of RTFO-aged PG 64-22, PG 64-22 + 0.5% W1 and PG 64-22 + 0.5% A1 are presented in Table 2. The contact angles of the binders were measured by testing the samples in a DCA apparatus using three probe liquids, namely water, glycerin and formamide. Generally, a contact angle of less than 90° indicates that the solvent can wet the surface of that material and vice versa (Buddhala et al. 2011). From Table 2, it can be observed that, the contact angles of PG 64-22 binder with water, glycerin and Formamide reduced due to the addition of 0.5% W1 additive. For example, the contact angle of RTFO-aged PG 64-22 binder was found to be 107.9°, 94.3° and 90.5° for water, glycerin and Formamide, respectively. The corresponding contact angles for the binder blend containing 0.5% W1 additive were 107.8°, 94.1° and 89.8°, respectively. A similar decreasing trend in contact angles was also observed with the addition of 0.5% A1 additive. The contact angles of PG 64-22 binder

Table 2. Contact angles of PG 64-22 binder blends

Binder type	Advancing contact angle (°)					
	Water		Glycerin		Formamide	
	Mean	Standard deviation	Mean	Standard deviation	Mean	Standard deviation
PG 64-22	107.9	0.09	94.3	0.1	90.5	0.11
PG 64-22 + 0.5% W1	107.8	0.14	94.1	0.32	89.8	0.34
PG 64-22 + 0.5% A1	107.0	0.08	94.0	0.13	89.7	0.29

Five asphalt binder samples were tested with each solvent

containing 0.5% A1 additive with water, glycerin and Formamide were found to be 107.0°, 94.0° and 87.9°, respectively. The reactions between the binder and the fatty amine derivatives of the W1 additive and polyamines of A1 antistripping agent are hypothesized to be responsible for the changes in contact angles. A similar trend of reduction in contact angle with an increase in the WMA and anti-stripping additive was reported in other studies (Buddhala et al. 2011; Wasiuddin et al. 2006, 2007; Ghabchi et al. 2013).

4.1.2 Surface Free Energy Components of Asphalt Binder

Typically, a change in the SFE components of a binder resulted in a change in the moisture-induced damage potential of asphalt-aggregate system. Bhasin et al. (2006) reported that the acid component of the asphalt acts as a scale factor in calculation of dry adhesive bond strength. Table 3 presents the SFE components of PG 64-22, PG 64-22 + 0.5% W1 and PG 64-22 + 0.5% A1. It was observed that the nonpolar Lifshitz-van der Waals (Γ^{LW}) and the total SFE (Γ^{Total}) component of PG 64-22 binder increased with the addition of the W1 additive. For example, the Γ^{LW} for PG 64-22 binder was found to increase from 9.31 to 10.40 mJ/m^2 upon the addition of 0.5% W1 additive. However, the acid (Γ^{+}) and base (Γ^{-}) components of PG 64-22 binder were found to reduce with the addition of W1 additive. The Γ^{+} was found to reduce from 1.81 to 1.51 mJ/m^2 and the Γ^{-} reduced from 0.68 to 0.64 mJ/m^2 for PG 64-22 + 0.5% W1 binder. Also, the Γ^{+}/Γ^{-} ratio of PG 64-22 binder was found to reduce from 2.66 to 2.36 with the addition of 0.5% W1 additive. This indicates that the asphalt binder may become more basic with the addition of W1 additive which may result in weak bonding with basic aggregates. Bhasin et al. (2006) tested several aggregates, namely granite, gravel, limestone, quartz and sandstone from different sources and found out that the Γ^{-} components of all the aggregates are relatively higher than the acid components. Therefore, the addition of W1 additive are expected to result in a weak adhesive bond with basic aggregates than the neat binder.

From Table 3, it was also observed that the Γ^{LW}, Γ^{Total} and the Γ^{-} components of PG 64-22 binder increased due to the addition of anti-striping additive. The Γ^{LW} and Γ^{-} were found to increase by approximately 11 and 22% with the addition of 0.5% antistripping additive, respectively. The Γ^{+} component of PG 64-22 binder was found

Table 3. SFE components of tested asphalt binders

Binder type	Surface free energy components (mJ/m^2)					
	Γ^+	Γ^-	Γ^{LW}	Γ^{AB}	Γ^{Total}	Γ^+/Γ^-
PG 64-22	1.81	0.68	9.31	2.22	11.54	2.66
PG 64-22 + 0.5% W1	1.51	0.64	10.40	2.00	12.43	2.36
PG 64-22 + 0.5% A1	1.49	0.83	10.32	2.22	12.57	1.80

Five asphalt binder samples were tested with each solvent

to reduce by approximately 18% with the addition of A1 antistripping agent. Also, the Γ^+/Γ^- component was found to reduce from 2.66 to 1.80 (approximately 32%) indicating more basic behavior with the addition of 0.5% antistripping additive. Same as contact angles, the reaction of the A1 antistripping agent with binder causes the changes in the binder surface energy properties.

4.2 XRF Tests Results of Asphalt Binder

Figure 2 presents the chemical compositions of PG 64-22, PG 64-22 + 0.5% W1 and PG 64-22 + 0.5% A1 binders from XRF tests. During XRF tests, all elements were scanned initially. Then those elements which were not detected, or spectrum could not be seen, were deleted. Each analysis was re-calculated with only the detected elements. Figure 2 presents the values of those elements which were detected. The results of the XRF tests indicate that oil constitutes more that 95% of the binder composition. Also, it was observed from the XRF spectrum that all three binders exhibited presence of Aluminium (Al, 1.49 keV), Silicon (Si, 1.74 keV), Sulfur (S, 2.31 keV), Chlorine (Cl, 2.62 keV),

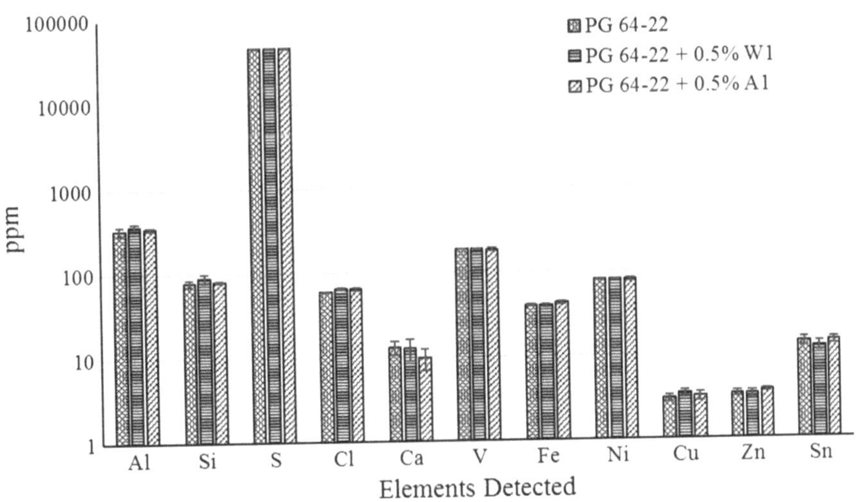

Fig. 2. Elements detected from XRF tests of PG 64-22, PG 64-22 + 0.5% W1 and PG 64-22 + 0.5% A1 binders (n = 3)

Calcium (Ca, 3.69 keV), Vanadium (V, 4.95 keV), Iron (Fe, 6.40 keV), Nickel (Ni, 7.48 keV), Copper (Cu, 8.05 keV), Zinc (Zn, 8.64 keV) and Tin (Sn, 25.27 keV). Hesp and Shurvell (2010, 2013) also mentioned the presence of the abovementioned elements in their tested asphalt binder. Among all the elements detected in PG 64-22 binder, the S was found to exhibit the highest proportion (46,507 ppm). Other elements can be listed as Al, V, Ni, Si, Cl, Fe, Sn, Zn and Cu according to their composition from highest to lowest. It was found that the Al, Si, S, Cl, V, and Cu exhibited an increase in the amount with an addition of 0.5% W1 additive. For example, the S was found to increase from 46,507 ppm to 47,192 ppm upon the addition of 0.5% W1 additive. However, Ca, Fe, Zn and Sn was found to decrease with the W1 additive. The changes may be resulted from the reaction of the chemical components of binder and the fatty amine derivatives of the W1 additive.

The Al, Si, S, Cl, Fe, Cu, Zn and Sn exhibited an increase in the composition with the addition of antistripping additive. The S was found to increase from 46,507 ppm to 46,722 ppm upon the addition of 0.5% A1 additive. However, Ca, V and Ni was found to decrease with the addition of A1 additive. The A1 additive is polyamine based antistripping agent. Same as WMA additive, the changes in the chemical compositions of the binder are expected to result from the reaction with the polyamines.

4.3 Correlation Between SFE Components and Chemical Constituents

Hefer et al. (2005) reported that the primary constituent of the binder is lightweight, oily or waxy fraction of long carbon chains and rings saturated with hydrogen. These oily fractions are generally nonpolar in character and made up of single C–H and C–C bonds, with relatively balanced electron distributions. These non-polar molecules interact through van der Waals forces and are expected to be responsible for the non-polar Lifshitz-van der Waals component (Γ^{LW}) the binder. Figure 3a presents the changes in the non-polar Lifshitz-van der Waals component (Γ^{LW}) with the oil percentage for the three tested binders. A good correlation (R^2 = 0.99) was observed between the Γ^{LW} and oil (%) of the binders.

Hefer et al. (2005) also mentioned that the polar molecules of the binders exhibit active sites. These active sites help to interact with other active sites within the binder (cohesive bonding) and with aggregate surfaces (adhesive bonding). Among the elements detected from XRF test, Al, Ca, Fe, Cu, Zn and Sn can exist in their positive ionic configuration and are expected to be responsible for positively active sites. Therefore, the relations between the Lewis acid component and the amount of Al, Ca, Fe, Cu, Zn and Sn in the binder blends were evaluated. The Lewis acid component was found to have very poor correlation with Ca, Fe, Zn and Sn. The Al and Cu exhibited better correlations with a R^2 values of 0.63 and 0.55. It indicates that the presence of Al and Cu might have higher influence on Lewis acid component than the other elements. Figure 3b presents the variation of Lewis acid component with Al. Similarly, the influence of Cl and S on the Lewis base component was evaluated. Both Cl and S exhibited very poor correlation with the Lewis base component. Figure 3c presents the variation of the Lewis base component with respect to S.

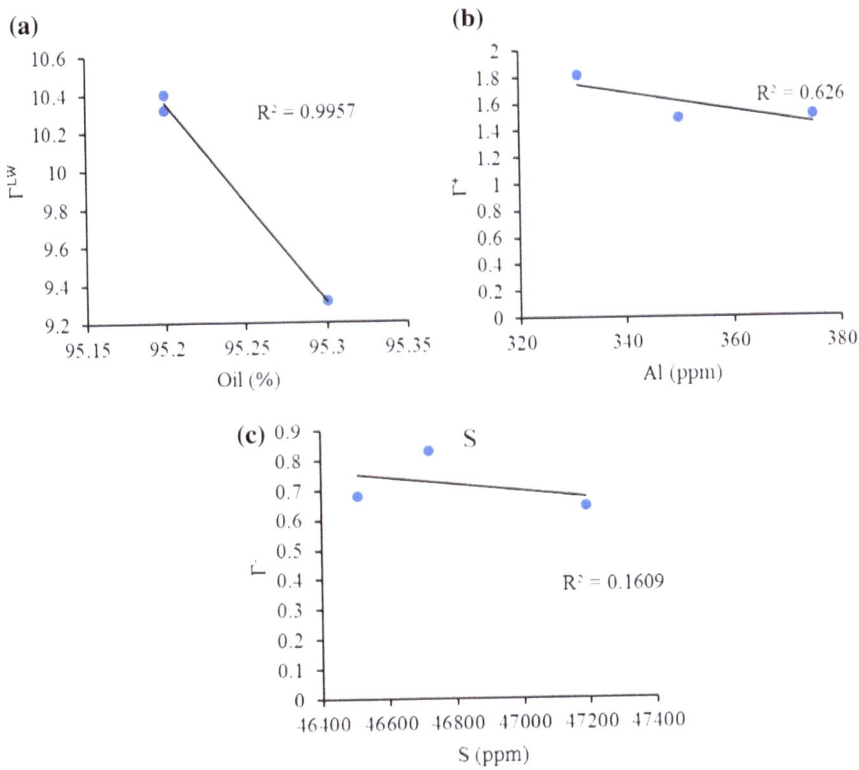

Fig. 3. Variation of **a** Γ^{LW} with oil percentage; **b** Γ^{+} with Al; and **c** Γ^{-} with S

5 Conclusions

In this study, a PG 64-22 binder was blended with a commercially available WMA additive and an antistripping additive. All the binders were tested for their Surface free energy characteristics using DWP test. The chemical characteristics of the same binders were also determined using XRF technique. Based on the results reported in the preceding sections, the following conclusions can be drawn:

i. The contact angles of PG 64-22 binder with water, glycerine and formamide were found to reduce with the addition of both the WMA additive and antistripping agent.
ii. The addition of WMA additive was found to increase the nonpolar Lifshitz-van der Waals component (Γ^{LW}) and total SFE component of the asphalt binder. However, the Γ^{+}/Γ^{-} of PG 64-22 binder was found to reduce with the addition of 0.5% WMA additive indicating a probability of weak bonding with basic aggregates.
iii. Addition of A1 antistripping additive to the asphalt binder was found to increase the Γ^{LW} and Γ^{-} components but reduce Γ^{Total} and Γ^{+} components of the binder.

iv. The XRF technique was successfully used to determine the changes in the chemical constituents of the binder due to the addition of a WMA and an antistripping additive.
v. The oil (%) of the binder blends were found to exhibit a good correlation with non-polar Lifshitz-van der Waals component. The amount of Al and Cu are expected to have significant influence on the Lewis acid component of the binders.

6 Recommendations for Future Study

The following recommendations could be made based on the limitations and the scope of the present study:

i. The chemical and surface energy properties of binders with different amounts of W1 and A1 additives should be determined to get a better understanding of the effects of additives.
ii. Further study is needed with XRF technique with different WMA and anti-stripping additives to understand the behavior of surface characteristics of asphalt binder.
iii. The effects of the addition of different additives on the SFE properties as well as moisture-induced damage potential of binders should also be evaluated.
iv. Future studies should focus on determining relationships between SARA fractions and SFE components of binders.

Acknowledgements. The authors gratefully acknowledge the financial support provided by the Oklahoma Department of Transportation (ODOT) and the Southern Plains Transportation Center (SPTC). The authors also extend their thanks to the binder suppliers, Ingevity and Arr-Maz for their support.

References

Arabani, M., Hamedi, G.H.: Using the surface free energy method to evaluate the effects of liquid antistrip additives on moisture sensitivity in hot mix asphalt. Int. J. Pavement Eng. ASCE **15** (1), 66–78 (2014)

Bagampadde, U., Isacsson, U., Kiggundu, B.M.: Impact of bitumen and aggregate composition on stripping in bituminous mixtures. Mater. Struct. **39**(3), 303–315 (2006)

Bhasin, A., Masad, E., Little, D., Lytton, R.: Limits on adhesive bond energy for improved resistance of hot-mix asphalt to moisture damage. Transp. Res. Rec. J. Transp. Res. Board **1970**, 3–13 (2006)

Bhasin, A., Little, D.N., Vasconcelos, K.L., Masad, E.: Surface free energy to identify moisture sensitivity of materials for asphalt mixes. Transp. Res. Rec. J. Transp. Res. Board **2001**, 37–45 (2007)

Buddhala, A., Hossain, Z., Wasiuddin, N.M., Zaman, M.: Effects of an amine anti-stripping agent on moisture susceptibility of sasobit and aspha-min mixes by surface free energy analysis. J. Test. Eval. **40**(1), 1–9 (2011)

Caro, S., Masad, E., Bhasin, A., Little, D.N.: Moisture susceptibility of asphalt mixtures, Part 1: Mechanisms. Int. J. Pavement Eng. **9**(2), 81–98 (2008)

D'Angelo, J.A., Harm, E.E., Bartoszek, J.C., Baumgardner, G.L., Corrigan, M.R., Cowsert, J.E., et al.: Warm-mix asphalt: European practice. FHWA-PL-08-007 (2008)

Hefer, A.W., Little, D.N., Lytton, R.L.: A synthesis of theories and mechanisms of bitumen-aggregate adhesion including recent advances in quantifying the effects of water. J. Assoc. Asph. Paving Technol. **74**, 139–196 (2005)

Hefer, A.W., Bhasin, A., Little, D.N.: Bitumen surface energy characterization using a contact angle approach. J. Mater. Civ. Eng. ASCE **18**(6), 759–767 (2006)

Hesp, S.A., Shurvell, H.F.: X-ray fluorescence detection of waste engine oil residue in asphalt and its effect on cracking in service. Int. J. Pavement Eng. **11**(6), 541–553 (2010)

Hesp, S.A., Shurvell, H.F.: . Quality assurance testing of asphalt containing waste engine oil. In: International Journal of Pavements Conference, São Paulo, Brazil (2013)

Hossain, Z., Lewis, S., Zaman, M., Buddhala, A., O'Rear, E.: Evaluation for warm-mix additive-modified asphalt binders using spectroscopy techniques. J. Mater. Civ. Eng. ASCE **25**(2), 149–159 (2012)

Ghabchi, R., Singh, D., Zaman, M.: Laboratory evaluation of stiffness, low-temperature cracking, rutting, moisture damage, and fatigue performance of WMA mixes. Road Mater. Pavement Des. **16**(2), 334–357 (2015)

Ghabchi, R., Singh, D., Zaman, M., Tian, Q.: Mechanistic evaluation of the effect of WMA additives on wettability and moisture susceptibility properties of asphalt mixes. J. Test. Eval. ASTM **41**(6), 1–10 (2013)

Gorkem, C., Sengoz, B.: Predicting stripping and moisture induced damage of asphalt concrete prepared with polymer modified bitumen and hydrated lime. Constr. Build. Mater. **23**(6), 2227–2236 (2009)

Le Guern, M., Chailleux, E., Farcas, F., Dreessen, S., Mabille, I.: Physico-chemical analysis of five hard bitumens: identification of chemical species and molecular organization before and after artificial aging. Fuel **89**(11), 3330–3339 (2010)

Lu, Q., Harvey, J.T.: Investigation of Conditions for Moisture Damage in Asphalt Concrete and Appropriate Laboratory Test Methods. Research Report No.: UCPRC-RR-2005-15, California Department of Transportation, Sacramento, CA 94273-0001 (2008)

Masad, E., Zollinger, C., Bulut, R., Little, D.N., Lytton, R.L.: Characterization of moisture damage using surface energy and fracture material properties. J. Assoc. Asphalt Paving Technol. **75**, 713–732 (2006)

Moghadas, N.F., Hamedi, G.H., Azarhoosh, A.R.: Use of surface free energy method to evaluate effect of hydrate lime on moisture damage in hot-mix asphalt. J. Mater. Civ. Eng. ASCE **25** (8), 1119–1126 (2012)

Mirzababaei, P.: Effect of zycotherm on moisture susceptibility of warm mix asphalt mixtures prepared with different aggregate types and gradations. Constr. Build. Mater. **116**, 403–412 (2016)

Soleimani, A., Walsh, S., Hesp, S.: Asphalt cement loss tangent as surrogate performance indicator for control of thermal cracking. Transp. Res. Rec. J. Transp. Res. Board **2126**, 39–46 (2009)

Van Oss, C.J., Chaudhury, M.K., Good, R.J.: Interfacial Lifshitz –van der Waals and polar interactions in macroscopic systems. Chem. Rev. **88**(6), 927–941 (1988)

Wasiuddin, N.M., Howell, D.C., Fogle, C.M., Zaman, M.M., O'Rear, E.A.: Acid-base characteristics of an asphalt binder with and without anti-strip additives. In: Airfield and Highway Pavement: Meeting Today's Challenges with Emerging Technologies, pp. 412–424 (2006)

Wasiuddin, N.M., Fogle, C.M., Zaman, M., O'Rear, E.A.: Effect of antistrip additives on surface free energy characteristics of asphalt binders for moisture-induced damage potential. ASTM J. Test. Eval. **35**(1), 36–44 (2007)
Wasiuddin, N.M., Zaman, M., O'Rear, E.A.: Effect of sasobit and aspha-min on wettability and adhesion between asphalt binders and aggregates. Transp. Res. Rec. **2051**, 80–89 (2008)
West, R., Rodezno, C., Julian, G., Prowell, B., Frank, B., Osborn, L.V., Kriech, T.: NCHRP Report 779: Field Performance of Warm Mix Asphalt Technologies. Transportation Research Board of the National Academies, Washington, DC (2014)

Implementing U.S. Freight Policy at the State Level: Oklahoma's Story

Dawn R. Sullivan(✉)

Oklahoma Department of Transportation, 200 N.E. 21st Street, Oklahoma City, OK 7301, USA
dsullivan@odot.org

Abstract. Safe and efficient transportation of freight is important to the U.S. economy and economic growth. Between 1975 and 1997 domestic intercity tons of the U.S. freight grew 60%, and that growth in freight movement was placing pressure on a congested highway system. Between 1980 and 2002, truck travel grew by 90% while lane miles grew only 5%. The percentage of urban interstates carrying 10,000 or more trucks was 27% in 1998, and forecasted to grow to 69% in 2020. Freight volumes were also expected to increase by 70% by 2020. In 2015, the U.S. Congress responded to the persistent and growing voice of freight stakeholders and made freight mobility a national issue in the Fixing America's Surface Transportation (FAST) Act. Each state is a key player to the implementation of freight mobility in the FAST Act. Oklahoma has a relatively large highway system mileage compared to its population, ranked 19th in state-owned system size. Located in the Crossroads of the U.S., Oklahoma is served by major highway corridors like Interstate 35, Interstate 40, and Interstate 44. The Oklahoma Department of Transportation (ODOT) is charged with planning, constructing, and maintaining Oklahoma's surface transportation infrastructure, including the interstate system, the U.S. highway system, and the Oklahoma highway system. ODOT began the process of developing a FAST Act compliant state freight plan in 2016. A Freight Advisory Committee (FAC) was created to assist in the planning process by helping to prioritize goals and identify concerns around particular operational issues such as bottlenecks. The Oklahoma Freight Transportation Plan (OFTP) was approved on December 4, 2017. This paper provides a critical overview of the OFTP relative to the conditions of Oklahoma's transportation infrastructure, particularly bridges and pavements. This plan sets forth a vision and goals, strategies and policies to achieve the goals, measures to track achievement, and investments selected because they support the goals. Importantly, ODOT has gone beyond planning for utilizing $100.2 million in National Highway Freight Program (NHFP) funds toward freight projects as required by the FAST Act. It has identified a series of investments for priority multimodal freight projects to be funded by traditional means, and all told has created an $875 million statewide freight investment program for the next five years.

Keywords: Freight movement · Policy · Transportation infrastructure Implementation · FAST Act

Z. Hossain et al. (eds.), *Solving Pavement and Construction Materials Problems with Innovative and Cutting-edge Technologies*, Sustainable Civil Infrastructures,
https://doi.org/10.1007/978-3-319-95792-0_15

1 Introduction

Transportation infrastructure funding in the United States is complex and varied as the 50 different states that make up the union. The way the U.S. funds highway transportation today was, for the most part, established by the Federal Aid Highway Act of 1973. Federal-Aid highway funds are authorized by Congress to assist the states in providing for construction, reconstruction, and improvement of highways and bridges on eligible Federal-Aid highway routes and for other special purpose programs and projects. The Federal-Aid Highway Program is a federally-assisted but state-operated program. The State Highway Agency (SHA) or State Department of Transportation (DOT) is the recipient of Federal funds and is also responsible for administering the Program. The role of the Federal Highway Administration (FHWA) is to administer the Federal-aid program in partnership with each state, with a vision of national goals and objectives, shared and implemented by the supporting interconnected State highway systems by providing financial assistance for the construction, maintenance and operations of the nation's 3.9 million-mile highway network, including the Interstate Highway System, primary highways and secondary local roads. The Federal Highway Administration (FHWA) is charged with implementing the Federal-aid Highway Program in cooperation with the States and local government.

Nearly 16 years ago, the Federal Highway Administration (FHWA) was looking into how transportation infrastructure and namely freight transport influenced the economy and helped with economic growth. The link between economic growth and transportation is tangible, yet notoriously difficult to quantify. In June of 2004, FHWA's Office of Policy Development published the report, Freight Transportation Improvements and the Economy; prepared by ICF Consulting and HLB Decision Economics, to research and understand the relationships between highway investment and economic goals.

The report noted that between 1975 and 1997 domestic intercity tons of Freight grew 60%, and that growth in freight movement was placing pressure on a congested highway system. Between 1980 and 2002, truck travel grew by 90% while lane miles grew 5%. The percentage of urban interstates carrying 10,000 or more trucks was 27% in 1998, and forecasted to grow to 69% in 2020. Freight volumes were also expected to increase by 70% by 2020.

The general findings and observations include the following:

- Freight transportation enhancements that reduce the costs of moving goods are critical for economic expansion.
- Investments that reduce the cost of moving goods to and from markets can help to increase and sustain economic growth.
- Efficiency and reliability of the freight transportation system affects economic productivity.

Research by Nadiri and Mamuneas (1996) cited in the report confirmed the role of transportation investments in economic growth. More and better roads reduce the cost of production in most industries, making it faster and cheaper to obtain parts and raw materials and to get finished products to market.

Fast forward to 2015, the U.S. Congress responded to the persistent and growing voice of freight stakeholders and addressed freight funding in the transportation reauthorization legislation. In the Fixing America's Surface Transportation (FAST) Act, freight mobility was specifically addressed and made freight a national issue. Each state is a key player as to how the United States plans for movement of freight, due in large part to the structure and administration of the federal aid highway program.

As previously stated, the Federal Aid Highway Program in the United States is a state administered federal program, with broad policy and regulatory requirements set at the federal level, but implemented by the states, reflecting local knowledge, priorities and realities. 2017 was the second year of a national freight program, and just now are the federally required state plans being formally adopted by the states and approved by FHWA, as required by the FAST Act (https://www.fhwa.dot.gov/fastact/).

The FAST Act requires that state goals be consistent with the national goals. National goals for freight are enumerated in the FAST Act and are summarized in Table 1.

Table 1. National freight goals

1.	Policies, operational improvements and investments for economic competitiveness; congestion and bottleneck reduction; reduced costs and improved year-round reliability; and productivity gain, especially by high-value job generators
2.	Safety, security, efficiency, and resilience—urban and rural
3.	Network state of good repair
4.	Economic efficiency and productivity of networks
5.	Improve short- and long-distance freight movement—across rural, rural-urban, and port/airport/gateway connections
6.	Flexibility for multistate corridor planning and organization
7.	Reduce environmental impacts
8.	Avoid burdens to state and local governments

Source WSP adapted from https://www.fhwa.dot.gov/fastact/legislation.cfm

2 The Oklahoma Story

Oklahoma is located in the South-Central plains of the United States and is characterized by a diverse and growing demographic and economic base. Major industries in Oklahoma include oil and gas, agriculture, aerospace, and manufacturing. The state's population in 2016 was 3.9 million, ranked 28th of the 50 states, and is projected to exceed 4.2 million in 2025. The population growth is expected to be strong in Oklahoma City and Tulsa, the state's two large metropolitan areas. Low to moderate growth is forecast in the remainder of the state (Long Range Transportation Plan 2015). Employment growth is forecast in much of the state. Freight miles of travel are expected to fall in line with U.S. Department of Transportation (U.S. DOT) projections and grow at a rate of slightly over 1% per year (https://www.rita.dot.gov/bts/press_releases/bts013_16, *Oklahoma from HIS*).

Oklahoma has a relatively large highway system mileage compared to its population, ranked 19th in state-owned system size. Oklahoma is served by major highway corridors like Interstate 35, that runs from the U.S. Mexican border in Laredo, Texas north to Duluth, Minnesota, Lake Superior and its border with Canada; Interstate 40, running from Barstow, California where it intersects with Interstate 15 the major north-south artery in southern California, east through the center of the U.S. across to Wilmington, North Carolina; Interstate 44 connecting to from Wichita Falls, Texas, diagonally the northeast across Oklahoma to St. Louis, Missouri where it connects to Interstate 70 and onward to Baltimore Maryland US-69/75. Arguably, Oklahoma is the crossroads of America.

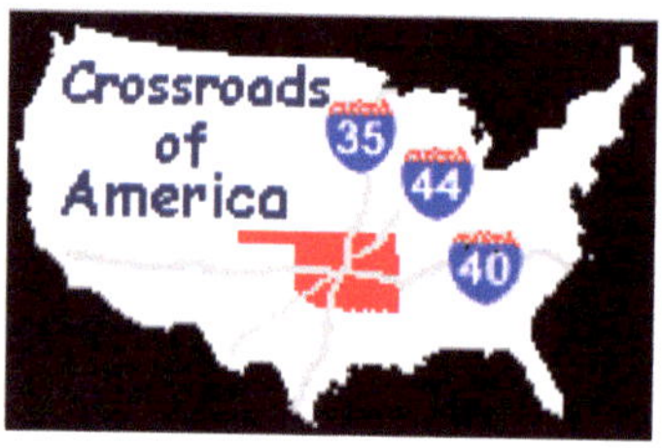

ODOT
Oklahoma Department of Transportation

The Oklahoma Department of Transportation (ODOT) is charged with planning, constructing, and maintaining Oklahoma's surface transportation infrastructure, including the interstate system, the U.S. highway system, and the Oklahoma highway system. ODOT also manages state-owned freight railroads, and administers other multimodal programs, including passenger rail, rural public transit, and the waterways program. ODOT is responsible for managing 12,265 centerline highway miles, 6800 bridges, with more than 70 million daily vehicle miles of travel.

ODOT began the process of developing a FAST Act compliant state freight plan in 2016. A Freight Advisory Committee (FAC) was created to assist in the planning process by helping to prioritize goals and identify concerns around particular operational issues such as bottlenecks. The FAC was important to sharing information related to industry, regulatory, and public priorities, and providing input on proposed strategies and projects. Members of the FAC included representatives from industries critical to the state's economy, representatives of transportation service providers, and multimodal facilities such as ports, Safety enforcement, planning organizations, tribal governments, and other state and federal agencies were also included. Four FAC meetings were held beginning in November of 2016 and extending through October of 2017. The Oklahoma Freight Transportation Plan was approved December 4, 2017 just in time for the legislative deadline.

Here are Oklahoma's freight goals in the priority order determined by the Oklahoma Freight Advisory Committee (FAC). Table 2 shows how freight goals correspond to an established Long Range Transportation Plan goal area and to establish national freight goals listed.

Freight is important to the transportation system and to the economy. Over 800-million tons of freight are transported annually in, out, within, and through Oklahoma. The value of goods transported annually is estimated at $1.3 billion (Oklahoma Long Range Transportation Plan 2015).

Table 2. Oklahoma's freight goals and correspondence to long-range transportation plan goals and national freight goals

LRTP goal area	OFTP freight goals	National freight goal #
Safe and secure travel	• Improve the safety and efficiency of freight movement and its interaction with other vehicles • Ensure the ability of urban and rural highways to safely accommodate growth in freight traffic	2
Infrastructure preservation	• Meet freight transportation needs by maintaining the Oklahoma State Highway System in a state of good repair • Support the preservation of Oklahoma multimodal freight networks through appropriate polices and initiatives	3, 5
Mobility: choice, connectivity and accessibility	• Ensure the competitive performance of the Oklahoma freight system • Foster a diverse portfolio of modal choices for Oklahoma's freight shippers and receivers in urban and rural areas • Support end-to-end operations of industry supply chains in Oklahoma markets for Oklahoma's industries	1, 4
Economic vitality	• Promote competitive access to domestic and international markets for Oklahoma's industries • Direct freight-related transportation investments to support the state's economy	1, 4
Environmental responsibility	• Support the growth of Oklahoma clean energy by promoting clean fuel use by freight providers • Avoid, minimize, or mitigate adverse environmental impacts related to freight transportation	7
Efficient intermodal system management and operation	• Capitalize on federal funding and finance programs to aid investment in the freight transportation system • Coordinate freight corridor development programs with neighboring states • Safeguard industry supply chains by improving resiliency of the freight transportation system to withstand disruptions	2, 6, 8

Source Oklahoma Freight Advisory Committee

The freight flow data presented here is based on profiles from the IHS Markit Transearch database, and supplemented with the FHWA's Freight Analysis Framework 4 (FAF 4) data. The latest year for which these historical data are available is 2014, and they were escalated to the base year of 2015 (Fig. 1).

Freight Flows by Highway, Railroad, and Waterway: 2011

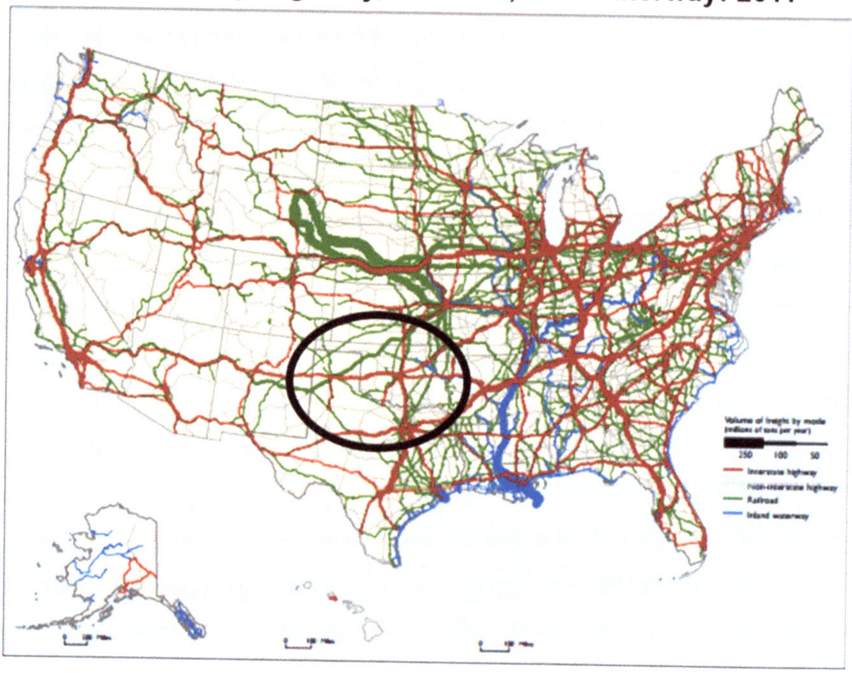

SOURCE: Highways: U.S. Department of Transportation, Federal Highway Administration, Freight Analysis Framework, Version 3.5, 2015; Rail: Based on Surface Transportation Board, Annual Carload Waybill Sample and rail freight flow assignments done by Oak Ridge National Laboratory; Inland Waterways: U.S. Army Corps of Engineers, Institute or Water Resources, Annual Vessel Operating Activity and Lock Performance Monitoring System data, September 2015.
BTS Freight Facts and Figures 2015

Fig. 1. Freight flows by highway, railroad, and waterway: 2011. https://www.rita.dot.gov/bts/sites/rita.dot.gov.bts/files/FFF_complete.pdf

Figure 2 shows total freight flows by direction (inbound, outbound, within state and through). Through freight relates to shipments that begin out-of-state, pass through Oklahoma, and continue to a destination out of the state. Note the large volume of pass-through freight.

Table 3 displays aggregate freight flows broken down by direction and by mode. Most of the volume of through tonnage is moved by rail.

Table 4 shows the growth in freight by tonnage between 2015 and 2045. Freight in Oklahoma is expected to grow by nearly 50% over the next 30 years. Most of this growth is projected to be in longer trips that have either an origin or destination point, or both, outside of the state. Through traffic is expected to see the greatest growth, at 63%.

In terms of modes, trucking will represent the largest mode share in 2045; however, rail freight is expected to grow at a slightly faster rate (nearly 48% compared to 45%) over the 30-year period. Water is expected to grow more slowly (35%).

Demand for freight rail service is expected to continue in Oklahoma, enhanced by the state's geographic location. Twenty-one freight railroads, including three Class I

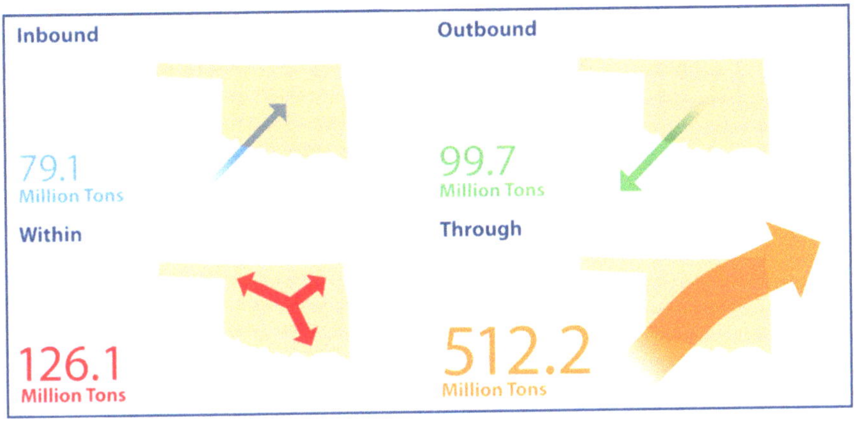

Fig. 2. Oklahoma freight flows (2015) by direction

Table 3. Oklahoma freight flows

Tonnage 2015 by mode and direction (millions)					
Mode	Inbound	Outbound	Within	Through	Total
Truck	46.5	78.5	123.6	224.3	472.9
Rail	29.5	18.0	2.5	287.9	337.9
Water	3.1	3.2	0.0	0.0	6.3
Total	79.1	99.7	126.1	512.2	817.1

Source IHS Transearch, WSP analysis, 2017

carriers, operate in the state. Attracting and training talented workers, and implementing new technology for safety and efficiency will continue to be important to the rail industry.

The lack of investment in infrastructure has resulted in highways, bridges and waterways that are obsolete and in disrepair. The U.S. freight railroads are private organizations that are responsible for their own maintenance and improvement projects. It is anticipated that railroad companies will need to continue adding to their systems to address the growth in rail traffic over the next decades. Once a source of pride and a great asset for U.S. businesses, many parts of the nation's transportation infrastructure network urgently require investment in renovation and restoration.

Nationally, the condition of the infrastructure has been of concern for many years. Table 5 demonstrates the bridge problem.

The overall condition of bridges in the U.S. has improved slowly over time. In 2000, 15.2% (89,415) of bridges were considered structurally deficient compared to 10.5% (63,521) in 2013.

Structurally deficient bridges are characterized by the deteriorated condition of bridge elements and reduced load-bearing capacity. In some cases, weight restrictions are placed on structurally deficient bridges, which may impact freight movement.

Table 4. Long-term Oklahoma freight growth (2015–2045)

Flow	Mode	Tons (Millions)		Percentage growth (%)
		2015	2045	2015–2045
Inbound	Truck	46.5	65.1	40.04
	Rail	29.5	29.7	0.60
	Water	3.1	3.9	25.38
	Total	79.1	98.7	24.76
Outbound	Truck	78.5	103.6	32.02
	Rail	18.0	24.0	32.83
	Water	3.2	4.6	45.01
	Total	99.7	132.2	32.57
Within	Truck	123.6	123.8	0.17
	Rail	2.5	3.2	30.99
	Total	126.1	127.0	0.78
Through	Truck	224.3	393.0	75.19
	Rail	287.9	441.7	53.42
	Total	512.2	834.7	62.95
Total	Truck	472.9	685.5	44.97
	Rail	337.9	498.6	47.54
	Water	6.3	8.5	35.14
	Total	817.1	1192.6	45.96

Source IHS Markit Transearch, Freight Analysis Framework 4.3, WSP analysis

Table 5. Condition of U.S. bridges 2013

Years old (as of 12/31/2013)						
	0–24	25–49	50–74	75–99	>99	All years
Total bridges	175,702	215,605	140,696	64,083	11,663	607,749
Total deficient bridges						
Number	18,680	41,231	49,646	30,445	7867	147,869
Percent	10.6	19.1	35.3	47.5	67.5	24.3
Structurally deficient						
Number	2576	16,200	22,491	17,388	4866	63,521
Percent	1.5	7.5	16.0	27.1	41.7	10.5
Functionally obsolete						
Number	16,104	25,031	27,155	13,057	3001	84,348
Percent	9.2	11.6	19.3	20.4	25.7	13.9

In the last decade, Oklahoma has focused the infrastructure funding that is available to ODOT on reducing the number of structurally deficient bridges and has made significant strides toward that goal. Oklahoma was proud to be recognized as number 1 in the nation in reducing the most structurally deficient bridges, down 2458 since 2007. Figure 3 demonstrates ODOT's progress. Still, the state of Oklahoma has significant challenges ahead as the bridge inventory continues to age, as shown in Fig. 4.

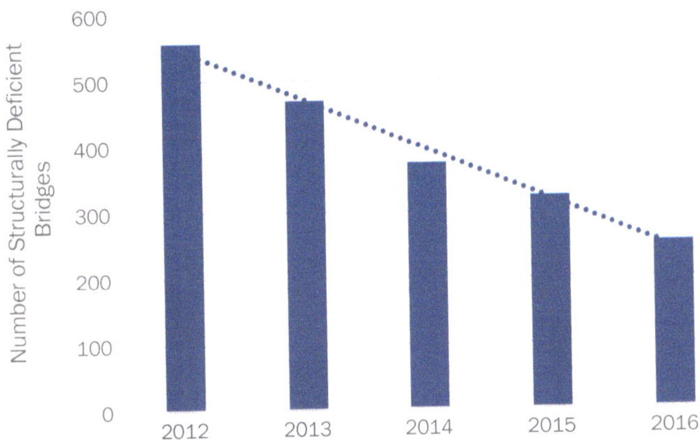

Fig. 3. Bridge condition. *Source* Oklahoma DOT

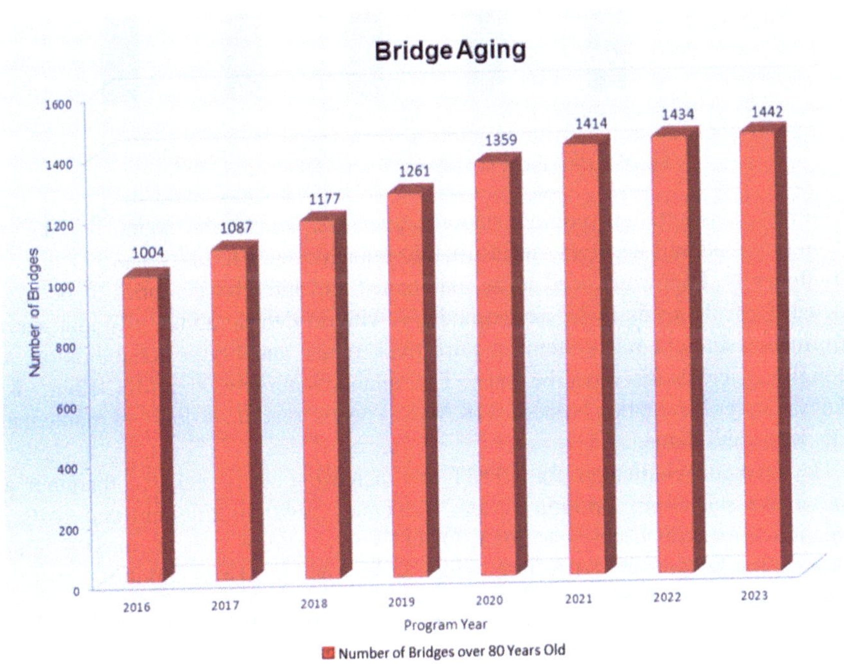

Fig. 4. Number of bridges in Oklahoma more than 80 years old, by year.*Source* Oklahoma DOT

3 Pavement Condition

For the most part, the major highways are in fair to good condition, with segments of the network northeast of Tulsa and in the Oklahoma Panhandle rated as poor. The challenge ODOT faces is to maintain pavement condition with current funding levels while addressing pavement deterioration.

Figure 5 displays the number of miles of pavement on the Oklahoma State Highway System that were rated poor throughout the state between 2012 and 2016, using the International Roughness Index (IRI). As shown, poor pavement mileage increases and decreases, but the overall trend is relatively flat.

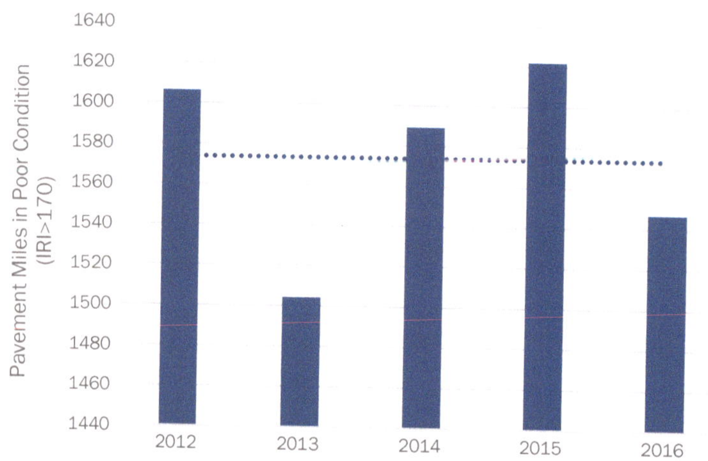

Fig. 5. Pavement condition. *Source* Oklahoma DOT

Freight transportation requires smooth pavement, structurally sound bridges, and ongoing railroad and waterway infrastructure improvements to deliver products safely and efficiently. Highways need to be maintained and interchanges need to be reconstructed. Growth needs to be accommodated without deterioration in freight service performance. Freight rail systems require track repair and bridge rehabilitation, and rail-highway crossings must be safe. The inland waterway, the McClellan –Kerr Arkansas River Navigation System (MKARNS) needs to address deferred maintenance on its locks and dams.

The financial challenges for ODOT are increasing as it seeks to maintain and improve the state transportation system. Revenue growth is minimal and costs are escalating. Based on fiscal year 2015, ODOT's total budget was about $1.9 billion, with a capital budget of $1.1 billion, which was broken down as follows:

- About 50% from federal revenue—mostly federal highway funds
- About 15% from state motor fuel tax
- The remaining 35% from a combination of bonds and other state and local revenues

With vehicle fuel efficiency increasing, and accelerating demands on the system, Oklahomans must address transportation funding issues. Oklahoma's 2015 through

2040 LRTP shows that the expected funding gap averages $360 million per year over 25 years, if current trends continue. Needs exceed expected available revenues by nearly 20% annually. Clearly a major component of addressing Oklahoma's freight needs is the challenge of finding additional funding.

Several important trends are likely to affect the demand for and availability of future freight transportation in Oklahoma:

- Energy independence will require increased production of crude oil. While this will be shipped principally by pipeline, sand required in the extraction process will be moved by rail.
- Agriculture will continue to be a growth industry consuming significant amounts of highway, rail, and waterway capacity.
- The changing retail trade environment will increase both urban and deliveries by truck; expanded number of branch distribution centers will also increase truck volumes; both will compete with through traffic for highway capacity.
- Technology advances supporting truck platoons could divert traffic from rail; safety concerns may require the construction of dedicated truck lanes, but also add to future congestion.

A major driver of the freight planning effort is the requirement of identifying the use of Oklahoma's apportionment of National Highway Freight Program (NHFP) funds in the state freight plan. The Plan considered various factors for the allocation of federal freight formula funds for Oklahoma's freight projects including level annual funding, corridor focus, geographic diversification, project ranking, stakeholder priorities, project size, and designation of critical candidate rural freight corridors. The result was the identification of 18 projects, selected to be funded in part with (NHFP) funds, constitute Oklahoma's *Five Year Financially Constrained Freight Investment Plan*. These projects require total funding of $250.5 million. NHFP funds will cover $100.2 million, and the remaining $150.3 million will be supplied by state and other federal sources. An additional $62 million (not included in the $250.5 million) already is being funded by a federal FASTLANE grant for U.S. 69 in Bryan County (Fig. 6).

U.S. 69 Bryan County

Visualization of US 69 after completion of FASTLANE grant project

Fig. 6. US 69 before and after completion of FASTLANE grant project

4 Additional Support by Traditional Federal and State Programs

In addition to projects funded in part by NHFP funds, 36 top highway freight mobility projects appear in the 8 Year Construction Work Plan. These projects are being funded from traditional highway sources, with 80% from the federal government and 20% from the state. These 36 projects represent an additional $504 million investment in freight over the next five years. Combined with the 18 projects that will receive NHFP funds, the total highway freight investment in Oklahoma over the next five years is $816 million.

5 Concluding Remarks

The Oklahoma freight transportation system serves the people of the state by delivering the necessities of everyday life: food, fuel, clothing, medicine, building materials and the equipment for communication, transportation, sporting and a multitude of other purposes. The system serves the businesses of Oklahoma by ensuring their supply lines, and giving them access to markets near and far, thus contributing to employment for people and prosperity for the state. To residents, these fundamental functions are largely invisible because they perform well, and their vital importance could attract attention only because of disruptive events. Even so, the quality of performance must be sustained at a favorable cost, so that Oklahoma is an affordable place to live and a competitive place for businesses to locate.

Good performance is reliable, productive, safe and secure; it is generated daily through freight operations and longer term through capital investments and policies in the public and private sectors. A high-quality transportation system benefits from multiple modes of transportation, because modal options keep competition sharp, thus influencing lower costs. A variety of modes accommodates a range of shipments whose volume, time commitments, and physical characteristics are quite diverse. The Oklahoma multimodal freight system does all these things. Moreover, it performs these functions for constituents well beyond its borders by means of the great quantities of goods that pass-through Oklahoma on the highways, railroads, and waterways of the state.

This is the first comprehensive freight plan ODOT has issued, although ODOT has considered the needs of freight in its transportation plans for many years. This Plan sets forth a vision and goals, strategies and policies to achieve the goals, measures to track achievement, and investments selected because they support the goals. Importantly, ODOT has gone beyond planning for utilizing $100.2 million in NHFP funds toward freight projects as required by the FAST Act. It has identified a further series of investments for priority multimodal freight projects to be funded by traditional means, and all told has created an $875 million statewide freight investment program for the next five years.

References

Nadiri, M.I., Mamuneas, T.P.: Contribution of highway capital to industry and national productivity growth. Report prepared for Federal Highway Administration Office of Policy Development, Work Order Number BAT-94-008 (1996)

Oklahoma Long Range Transportation Plan 2015–2040: Moving Oklahoma Forward. Prepared by Oklahoma Department of Transportation (2015)

Oklahoma Freight Transportation Plan 2018–2022: Prepared by Oklahoma Department of Transportation (2017)

https://www.fhwa.dot.gov/fastact/

https://www.rita.dot.gov/bts/press_releases/bts013_16

Author Index

Z. Hossain et al. (eds.), *Solving Pavement and Construction Materials Problems with Innovative and Cutting-edge Technologies*, Sustainable Civil Infrastructures,
https://doi.org/10.1007/978-3-319-95792-0

MIX
Papier aus verantwortungsvollen Quellen
Paper from responsible sources
FSC® C105338

If you have any concerns about our products,
you can contact us on
ProductSafety@springernature.com

In case Publisher is established outside the EU,
the EU authorized representative is:
Springer Nature Customer Service Center GmbH
Europaplatz 3, 69115 Heidelberg, Germany

Printed by Libri Plureos GmbH
in Hamburg, Germany